Marginalized Reproduction

Science in Society Series

Series Editor: Steve Rayner
James Martin Institute, University of Oxford
Editorial Board: Gary Kass, Anne Kerr, Melissa Leach, Angela Liberatore,
Jez Littlewood, Stan Metcalfe, Paul Nightingale, Timothy O'Riordan,
Nick Pidgeon, Ortwin Renn, Dan Sarewitz, Andrew Webster, James Wilsdon,
Steve Yearley

Business Planning for Turbulent Times

New Methods for Applying Scenarios
Edited by Rafael Ramírez, John W. Selsky and Kees van der Heijden

Democratizing Technology

Risk, Responsibility and the Regulation of Chemicals
Anne Chapman

Genomics and Society

Legal, Ethical and Social Dimensions
Edited by George Gaskell and Martin W. Bauer

Marginalized Reproduction

Ethnicity, Infertility and Reproductive Technologies
Edited by Lorraine Culley, Nicky Hudson and Floor van Rooij

Nanotechnology

Risk, Ethics and Law
Edited by Geoffrey Hunt and Michael Mehta

Unnatural Selection

The Challenges of Engineering Tomorrow's People
Edited by Peter Healey and Steve Rayner

Vaccine Anxieties

Global Science, Child Health and Society
Melissa Leach and James Fairhead

A Web of Prevention

Biological Weapons, Life Sciences and the Governance of Research
Edited by Brian Rappert and Caitríona McLeish

Marginalized Reproduction

Ethnicity, Infertility and Reproductive Technologies

Edited by
Lorraine Culley, Nicky Hudson and Floor van Rooij

from Routledge

First published by Earthscan in the UK and USA in 2009

Copyright © Lorraine Culley, Nicky Hudson and Floortje Barbera van Rooij, 2009

ISBN: 978-1-84407-576-8 (hbk)
ISBN: 978-0-41584-942-5 (pbk)

Typeset by FiSH Books, Enfield
Cover design by Susanne Harris

For a full list of publications please contact:

Earthscan
2 Park Square Milton Park Abingdon Oxon OX14 4RN
711 Third Avenue New York NY 10017

First issued in paperback 2013

A catalogue record for this book is available from the British Library

Library of Congress Cataloging-in-Publication Data
Marginalized reproduction : ethnicity, infertility and reproductive technologies /
edited by Lorraine Culley, Nicky Hudson and Floor van Rooij.
    p. cm.
    Includes bibliographical references and index.
    ISBN 978-1-84407-576-8 (hardback)
    1. Infertility—Cross-cultural studies. 2. Human reproductive technology—
Cross-cultural studies. I. Culley, Lorraine. II. Hudson, Nicky. III. Van Rooij,
Floor.
RC889.M365 2009
616.6'92–dc22

2008044793

Contents

List of Contributors

Karl Atkin is a Senior Lecturer in the Department of Health Sciences, University of York. He was previously Director of Primary Care Research at the University of Leeds. Karl is a medical sociologist, whose current research focuses on community engagement and genetic conditions, the social consequences of encephalitis, the experience of fertility impairment among people who have cancer, and the social exclusion of minority ethnic populations. He has published widely on the experience of family carers, young people and identity, disability and chronic illness, and the organization of health and social care.

Theda Borde is Professor of Social Medicine and Social Work at the Alice Salomon University of Applied Sciences in Berlin. She studied political science and public health, was head of an international education and counselling centre for migrants in Berlin, and has conducted research on migration and health in cooperation with the Charité Berlin. Her research has included the healthcare situation of migrant women in gynaecology, patient orientation in health and social care, the utilization of emergency rooms and transcultural aspects of the menopause. From 2002 to 2004 she was chairperson of the 'Committee of Experts on Health Services in a Multicultural Society' of the Council of Europe.

Rosario Ceballo is an Associate Professor of Psychology and Women's Studies at the University of Michigan. She earned her BA at Yale University in 1987, and she received a graduate certificate in Women's Studies in 1993 and a PhD in Clinical and Developmental Psychology in 1995 from the University of Michigan. Her research focuses on the impact of infertility and racial stereotypes about women's reproductive abilities on racial and ethnic minority women. Her broad research interests and publications also examine how contextual aspects of poverty, such as financial strain and community violence exposure, influence parenting and adolescents' academic and psychological functioning.

Noëlle Cotter is a PhD candidate in the School of Social Work and Social Policy in Trinity College, Dublin. Her doctoral research is a longitudinal study of couples' experiences of infertility and its treatment in Ireland and is funded by Ireland's Health Research Board. In addition to research in the area of reproductive technologies, her research experience includes work on environmentalism, racism, health and housing. Noëlle holds a BA in History and an MSc in Applied

Social Research from Trinity College, Dublin and is currently working as a researcher in the area of social and affordable housing.

Lorraine Culley is Professor of Social Science and Health and Associate Director of the Mary Seacole Research Centre at De Montfort University, Leicester, UK. She has recently completed with colleagues a major NHS-funded research project exploring the experience of involuntary childlessness in British South Asian communities and an Economic and Social Research Council (ESRC) funded project investigating public perceptions of third party assisted conception. Her current research work is focused on ethnicity and health and includes work on cultural diversity and endometriosis (funded by the UK Department of Health) and transnational aspects of assisted reproductive technologies (ARTs).

Yasmin Gunaratnam is a Lecturer in the Sociology Department at Goldsmiths College, University of London. She has been working on issues of race and gender equality in health and social care for the past 20 years. Yasmin has a particular interest in qualitative research methods and is author of *Researching 'Race' and Ethnicity: Methods, Knowledge and Power* (Sage, 2003). Yasmin has also jointly edited a book with David Oliviere entitled *Narrative and Stories in Health Care: Illness, Dying and Bereavement* (Oxford University Press, forthcoming).

Zeynep Gürtin-Broadbent is a PhD student in the Centre for Family Research, University of Cambridge. In her doctoral thesis she examines how the 'global' in vitro fertilization (IVF) technology has been localized in the Turkish context and explores Turkish women's experiences of infertility and assisted reproduction. Her research interests are broadly in gender, reproduction, reproductive medicine and its delivery, patient–clinician interactions, bodies, and reproductive decision making, and she has a particular area interest in Turkey. She works in an interdisciplinary way, utilizing a range of qualitative research methods, principally ethnography and in-depth interviews. Zeynep is co-Convener of the Cambridge Interdisciplinary Reproduction Forum.

Nicky Hudson is a Senior Research Fellow in the School of Applied Social Sciences at De Montfort University, UK. She is a sociologist with an interest in health, medicine, gender and ethnicity, and recently completed her doctoral research: 'Infertility in British South Asian communities: Negotiating the community and the clinic', which explored the experiences of British South Asian women who have used ARTs. Nicky is also interested in qualitative research methodology, particularly in aspects of the researcher–researched relationship.

Marcia C. Inhorn is the William K. Lanman Jr Professor of Anthropology and International Affairs in the Department of Anthropology and The Whitney and Betty MacMillan Center for International and Area Studies at Yale University. She also serves as Chair of the Council on Middle East Studies there. A specialist on Middle Eastern gender and health issues, Marcia has conducted research on the social impact of infertility and ARTs in Egypt, Lebanon, the United Arab Emirates and Arab America over the past 20 years. She is the author of three books on the subject, as well as co-editor of six books in the areas of gender, health, science, technology, biomedicine and public health.

Mark R. D. Johnson is Professor of Diversity in Health and Social Care, and directs the Mary Seacole Research Centre, De Montfort University, Leicester, UK. His recent publications include articles on visual impairment, health promotion, research ethics, alcohol use among 'second generation' migrants, community cohesion, and health services for asylum seekers, as well as books, and reports for the UK Home Office and Department of Health on aspects of health inequality, 'race' and ethnicity. He is also actively working with community groups on healthcare issues such as innovative ways to raise uptake of services for the visually impaired, and to validate translated health promotion materials.

Dineke Korfker is a midwife and anthropologist. Since 2001, she has worked at TNO Quality of Life, Section of Reproduction and Perinatology. TNO is a knowledge organization that makes scientific knowledge applicable for companies, government bodies and public organizations. Her main research interests are reproductive health, migrants and domestic violence, with a special focus on violence during pregnancy and female genital mutilation. Previously, she worked for many years in Africa (Mozambique and Egypt) for the United Nations Population Fund (UNFPA) as an adviser in reproductive health programmes.

Pranee Liamputtong has a Personal Chair in Public Health at La Trobe University, Australia. She is a medical anthropologist and has interests in the health of immigrants and refugees, women's health, mothers' and children's health, and reproductive and sexual health. Pranee has carried out a number of research projects with immigrant women in Australia and women in Southeast Asia. She has recently undertaken qualitative research with women living with HIV/AIDS in Thailand. Pranee has written a number of text books including *Researching the Vulnerable: A Guide to Sensitive Research Methods* (Sage, 2006) and *Doing Cross-Cultural Research: Ethical and Methodological Perspectives* (Springer, 2008).

Robert Nachtigall is a Reproductive Endocrinologist and Clinical Professor of Obstetrics, Gynecology and Reproductive Sciences at the University of California, San Francisco whose research career has been committed to the study of the ethical and psychosocial issues and controversies that accompany many aspects of modern infertility treatment. His work has been funded by the National Institutes of Health and the National Institute on Aging and includes an investigation of how infertile couples who have undergone IVF decide what to do with their surplus frozen embryos, an ethnographic qualitative description of the infertility experiences of low-income Latino women, and an exploration of the process that parents who conceive with donor gametes engage in as they decide whether or not to inform their children of the true nature of their conception.

Frank van Balen studied sociology, political science, macro-economics and non-Western sociology. He is Associate Professor at the Faculty of Social and Behavioural Sciences at the University of Amsterdam and was, until retirement, programme director of the Section Education and Child-Rearing. He carried out and supervised various major research projects in the field of involuntary childlessness, children conceived with reproductive technologies, same-sex families, prenatal diagnostics, sex selection, the cross-cultural value of children, and infertility and stress. He is President of the Dutch Society for Social Research on Infertility (SORT). Frank van Balen co-edited the volume *Infertility around the Globe* (University of California Press, 2002), which received the 'best volume' award from the American Anthropological Association. He is initiator of a global network of social scientists in infertility research, and co-organized the first conference in this field. Frank retired in 2008, but still continues to do research and supervision.

Floor van Rooij recently obtained her PhD at the Faculty of Social and Behavioural Sciences of the University of Amsterdam, where she conducted a study entitled 'The experience of involuntarily childless Turkish migrants in the Netherlands: Parenthood motives, psychosocial consequences, responses and help-seeking behavior'. Her research interests include involuntary childlessness, ethnic minorities, healthcare and parenting. She is currently working as a researcher at the Department of Epidemiology, Documentation and Health Promotion in the Public Health Service of Amsterdam (GGD Amsterdam).

Foreword

Throughout the world, infertility – the inability to have desired children – is stigmatized and viewed by fertile and infertile alike as a failure to achieve important cultural goals. Because the concern with fertility is so intimately involved with other cultural institutions, interpretations, treatments and the social consequences of infertility will vary from one sociocultural setting to another. For this reason, the social scientific study of infertility is not to be regarded simply as a practical enterprise oriented to providing better support to the infertile and to those who treat them, but also as a window through which to observe sociocultural reality.

For example, infertility provides us with a lens on gender. The way people respond to infertility tells us much about the roles available to women, the relative power of men and women, and the texture of the relationship between women and men within the family, kinship networks and other sociocultural settings. Infertility also presents us with an opportunity to observe medical institutions in action. In industrialized societies, infertility has been medicalized and medical institutions often exert hegemony over how many people experience infertility, and we can look at infertility as a site where individuals adapt to and confront the world of high-tech medicine. In addition, infertility presents us with the opportunity to observe the quest for meaning in contemporary societies. Whenever people are confronted with a failure to achieve desired goals, they look for solutions at the level of meaning. Infertility is thus an ideal site for watching people ascribe meaning to events that the medical model treats as scientific and therefore morally neutral.

This book – *Marginalized Reproduction: Ethnicity, Infertility and Reproductive Technologies*, by Lorraine Culley, Nicky Hudson and Floor van Rooij – draws our attention to two other features of contemporary societies that appear in sharper focus when viewed through the lens of infertility. People tend to refer to both of these features under the rubric of 'ethnicity', but I think it is helpful to distinguish between ethnic stratification and cultural pluralism. Studying infertility gives us a way to appreciate at a very personal level the tension between the stratification that pervades all industrialized societies and the ideology of equal citizenship that industrialized states articulate. At the same time infertility can serve as a vantage point from which to watch the processes by which migrants struggle to adapt their cultural reality to the new realities that surround them and by which host societies struggle to develop a new conception of national identity.

There are now a number of studies of the experience of infertility in industrialized societies, but these studies – including my own – have highlighted the perspectives of White, middle-class treatment seekers. There are also a number of studies of the experience of infertility in resource-poor 'developing' societies. Comparing these sets of studies, we can discern the existence of 'two worlds' of infertility. In one world, the biomedical model exercises hegemony, medical solutions to infertility are available to many and viable alternatives to motherhood exist. In the other world, the biomedical model coexists with or competes with more holistic conceptions of health and well-being, access to medical care is more limited and alternatives to motherhood seem less viable.

For the first time, in this volume, we have a collection of chapters that describe what happens when the two worlds of infertility come together, as they do in all industrialized societies. Here we get to see that global reproductive stratification exists within as well as between societies. Here, too, we have the opportunity to observe the way individuals navigate between competing realities and operate in the context of limited opportunities. The first half of the book provides an overview of theory research, and methodological challenges involved in studying infertility among marginalized people. The second half features case studies which highlight commonalities and differences in the experience of infertility. The editors have done an excellent job of compiling in one place a group of informative and interesting chapters that draw our attention to a new perspective from which to view both the experience of infertility and the reality of industrialized societies. They are to be commended for the service they have performed for the scholarly community.

Arthur L. Greil
Alfred University
New York
January 2009

Acknowledgements

We thank our contributors for agreeing to be part of this collection. The project developed out of a colloquium on Ethnicity and Reproductive Technologies held in Nottingham, UK in September 2006. Funding for this colloquium came from the Economic and Social Research Council, Science in Society Programme, (related to Grant No RES-160-25-0044) with assistance from the Society of Reproductive and Infant Psychology and we would like to thank the funders for their sponsorship of this event. We are also grateful to our colleagues who took part in the colloquium for their encouragement and contribution to our debates on many of the issues discussed in this book. Finally we would like to thank all the research participants who have contributed to the studies discussed in this collection, especially the many women and men who have so generously shared with our contributors, the stories of their experience of infertility.

List of Acronyms and Abbreviations

ARTs	assisted reproductive technologies
BICA	British Infertility Counselling Association
BME	black and minority ethnic
BMI	body mass index
CAHR	Commission on Assisted Human Reproduction
ESRC	Economic and Social Research Council
GP	general practitioner
HFE	Human Fertilisation and Embryology (Act)
HFEA	Human Fertilisation and Embryology Authority
ICSI	intracytoplasmic sperm injection
IUI	intrauterine insemination
IVF	in vitro fertilization
NHS	National Health Service
NIH	National Institutes of Health
NRTs	new reproductive technologies
PCOS	Polycystic Ovary Syndrome
STD	sexually transmitted disease
UNFPA	United Nations Population Fund

Introduction: Ethnicity, Infertility and Assisted Reproductive Technologies

Lorraine Culley, Nicky Hudson and Floor van Rooij

Infertility is a common occurrence. Differences in definitions, measurement criteria and healthcare systems between countries make global estimates of the prevalence of infertility difficult. However, a recent and comprehensive review of 25 population surveys of infertility concluded that overall around 9 per cent of women aged 20–44 experience infertility, which equates to 72.4 million women worldwide (Boivin et al, 2007).[1] The inability to conceive has been documented in an extensive collection of studies as an experience that has a profound influence on the personal well-being of women and men (Phoenix et al, 1991; Monach, 1993; Sandelowski, 1993; Whiteford and Gonzalez, 1995; Inhorn, 1994, 1996; Franklin, 1997; Letherby, 1999; Becker, 2000; Reissman, 2000; Throsby, 2004; Allan, 2007). Guilt, helplessness, marital stress and depression are commonly reported, particularly for women, although as Greil's (1997) review of the social psychological literature argues, there are many methodological flaws in existing studies, including small sample sizes and over-reliance on treatment seekers.[2] Boivin et al (2007) estimate that only around half of those experiencing fertility problems seek any infertility care, and even in more developed societies less than one quarter actually receive any specialist fertility treatment.

Nevertheless, the development of assisted reproductive technologies (ARTs) such as in vitro fertilization (IVF) has meant that many infertile couples can be helped to achieve a pregnancy, although for all fertility treatments there is a less than 50 per cent chance of successful conception and many treatments have potentially harmful side effects (van den Akker, 2002).[3] Since the birth of the first baby using IVF in 1978 more than 3 million babies have been born using ART, with an estimated 1 million ART cycles now performed each year, producing around 200,000 babies worldwide (ESRHE, 2006). The range of treatments has expanded substantially in the last 30 years. In particular, the advent of intracytoplasmic sperm injection (ICSI) has radically changed the potential for 'treating' male infertility, providing a genetic link to the offspring for sub-fertile men. For those who can afford this procedure, this has largely replaced donor insemination in Western societies and is growing in popularity in other parts of the world (Inhorn, 2003). Many relatively 'low-tech' treatments (drug therapy and intrauterine insemination (IUI)) are still used for some categories of infertility, especially as first-stage treatment for younger women, and as the only accessible form of treatment for those who cannot afford the expensive option of IVF.

Data on international treatment cycles highlight the ways in which ARTs are unequally distributed globally. Only a few countries offer access to IVF and ICSI. Of the 91 member states of the World Health Organization (WHO), 48 had medical facilities offering IVF in 2000 (Natchigall, 2006). Half of all cycles were delivered in just four countries (US, Germany, France and UK).[4] Despite all this activity in ART, it is salutary to remember that high-tech, high cost (and in many cases high profit) IVF delivers a live birth rate of less than 25 per cent per cycle (Ombelet et al, 2008).[5]

While there is a growing volume of literature on the use of ARTs in Western societies, the experience of infertility in minority ethnic communities remains largely invisible to epidemiologists, social scientists, practitioners and policy makers. There is limited research that explores the impact of diverse ethnicities, religious identities and cultural contexts on the experience and resolution of infertility within the West. This book attempts to address this research lacuna, demonstrating the potential importance of the social and cultural context of infertility and its treatment and, in several chapters, foregrounding the experiences of marginalized and racialized minorities within the West.[6] Drawn from a range of disciplinary perspectives, theoretical frameworks and geographical locations, the collection brings together for the first time work that has itself been somewhat on the margins of the academy.

ARTs and social science

Infertility has attracted the attention of a growing and diverse constituency of scholars from the biological, behavioural and social sciences, as well as cultural critics, ethicists, theologians and legal experts. This reflects, to some degree, its utility as a 'rhetorical vehicle' for a variety of Western concerns (Sandelowski and de Lacey, 2002). Part of the fascination of social scientists with infertility and its treatment is due to the intimate connections between reproduction and power. Studying reproduction provides a useful lens to explore the 'complex social arrangements through which legacies of property, positions, rights, and values are negotiated over time' (Ginsberg and Rapp, 1995, p2). Infertility as a form of 'reproduction gone awry' (Jenkins and Inhorn, 2003) has been a frame for exploring a number of concerns emanating from modern and post-modern social theory.

Thompson (2005), for example, in exploring the complex 'ontological choreography' of making babies and making parents using assisted reproduction, demonstrates the 'dynamic coordination of the technical, scientific, kinship, gender, emotional, legal, political, and financial aspects of ART clinics'; a bringing together of 'things that are generally considered parts of different ontological orders (part of nature, part of the self, part of society)' (p8). The rapid extension of reproductive technologies has deeply unsettled established ideas of what is 'natural' (Strathern, 1992; Franklin, 1997) and given rise to a number of new (and for many, unnerving) family forms and relationships. It is not surprising then, that infertility and assisted conception have generated considerable academic (and popular) interest.

The vast majority of this work, however, has concerned itself with the needs, interests and experiences of dominant social groups and white middle class couples in the West in particular. There are of, course, notable exceptions. The framing of infertility in terms of the global politics of reproduction, pioneered by feminist scholars for example, marks a considerable improvement in our understanding of infertility (Thompson, 2002). Extending their gaze beyond the traditional foci of anthropology, Ginsburg and Rapp (1995) alerted us, in a pioneering volume, to the importance of the politics of reproduction, locally and globally. They employ the term 'stratified reproduction' to explore how reproduction is structured across social and cultural boundaries, describing the ways in which power relations empower some categories of people to nurture and reproduce, while others are disempowered.

Global concerns

The challenge to the dominance of infertility as a Western phenomenon has been most richly portrayed in a groundbreaking text published in 2002. *Infertility around the Globe* (Inhorn and van Balen, 2002) explicitly focused on the global dimensions of infertility, moving beyond Western sites of technology production and debate, to expose the cross-cultural prevalence of infertility and the global connections between societies that produce and consume technologies which both enhance and curtail fertility. The rationale for this collection was a pressing need to reveal the 'lived experience' of infertility and childlessness, not just in the West, but in a range of non-Western societies.

This text played a vital role in decentring Western perspectives of infertility and technological solutions to childlessness, demonstrating the cultural variability in beliefs about infertility, its causes and consequences and in ideas about what should be done about it. The book was a powerful antidote to the privileging of Western concerns and concepts. Inhorn and van Balen foregrounded the paradox of infertility in 'high-fertility' cultures, challenging the scholarly silence in the West on the plight of the infertile in 'other' places. A silence which, they argued, reflects a common obsession in international population discourse, with curbing the 'hyperfertility' of non-Western subjects and a lack of desire to address the needs of infertile populations within non-Western settings (Inhorn and van Balen, 2002). This work, and other important contributions, have pointed to global contrasts in the consequences of infertility, with typically much more severe social consequences described for those in less developed societies (especially women) (Ombelet et al, 2008).

Global interaction in the development and application of ARTs has also been highlighted in the work of post-colonial/transnational feminist scholars who are concerned to highlight the potential for exploitation of the women of the 'South' which the new 'red biotechnologies' engender (Widdows et al, 2003; Dickenson 2004; Darling, 2006; Mulay and Gibson, 2006).[7]

Infertility and minorities in the West

Attention to globally 'stratified reproduction' (Ginsberg and Rapp, 1995) is most welcome. Our interest, in the present collection, however, turns to an alternative (though closely related) formulation of 'stratified reproduction', through a consideration of the experiences of infertile women and men from what are often marginalized communities within the West. The book represents a preliminary contribution to our concern to problematize the 'whiteness' and ethnocentricity of infertility research in the West, and to open up the field of race/ethnicity, culture and infertility. In doing so it brings together discussions of methodological challenges of research in the highly politicized arena of ethnic differentiation, and studies which have addressed the diverse contexts in which infertility is experienced, and new reproductive technologies accessed in Western societies.

The whiteness of infertility research is a consequence, at least in part, of the fact that in most Western countries, non-white groups are more likely to be among the 'have-nots' of ARTs (Thompson, 2002) and less likely therefore to be eligible for participation in such research. The marginalization of racialized minorities from biomedical citizenship is perhaps most clear in the US. The absence of concern about infertility among low-income women and 'women of color' has been noted as a serious social and political problem (Nsiah-Jefferson and Hall, 1989). This erasure reflects socio-economic disadvantage in a country where many poor people are priced out of healthcare. It also reveals, some argue, the workings of the legacy of slavery and a contemporary racism which positions 'women of color' as 'breeders' rather than mothers (Hartouni, 1994; Roberts, 1997).[8]

Drawing on primary research from the US, the UK, Eire, the Netherlands and Australia, this book turns the spotlight onto the specific ways in which minority status and cultural and religious mores might impact on the experience of infertility. Some of the chapters are concerned with the relative exclusion of racialized minorities from treatment. Others explore the particular ways in which culture and the broader social context might impact on the experience of infertility and its treatment. A significant proportion of the book is given over to the discussion of important methodological questions. What particular challenges face researchers investigating ethnicity and infertility? How should we capture ethnicity data and how can this be compared across different nation states? How does the identity of the researcher impact on the research process?

A central concern of the book is to explore how ethnicity and culture might shape the experience of infertility in the West. It is predicated on the view that infertility is not simply a medically diagnosed reproductive impairment, but a socially constructed reality (Greil, 1991). The most important decisions that people make – to define themselves as infertile, to seek treatment (or not), to decide between forms of treatment, to consider adoption or other alternatives and so on – are all part of a fundamentally social process. Furthermore, as several chapters in this volume show, the ability to access biomedicine and the quality of the care received are themselves also conditioned by the social and political processes of marginalization. Infertility and new reproductive technologies, then, will be differentially experienced according to an individual's or couple's location in a social space, and their agency in negotiating and navigating that location.

Of course the interrogation of the significance of ethnicity and culture in infertility in the West raises the contentious issue of how to conceptualize culture(s) and the impact of culture on health and healthcare (Ahmad, 1996; Kelleher, 1996; Culley, 2006). Several chapters in this volume challenge the kind of cultural theorizing that somewhat hastily invokes culture when faced with anything we cannot otherwise understand, often overlooking alternative explanations (Fassin, 2001). Cultural essentialism is rife in public health research. Ethnicity is commonly conceived as something possessed only by non-Western or minority groups. The 'Other' is seen as uniquely determined by the dictates of culture, propelled to think and act in particular ways (Ahmad, 1996). The reification of culture plays down heterogeneity within cultures, exaggerates lines of demarcation between different cultural groups, affords culture too much explanatory weight and imagines the cultural 'other' as devoid of agency (Phillips, 2007). Yet, the de-construction or de-essentializing of culture in theories of hybridity, diaspora and negotiated identities (e.g. Hall, 1992; Bhabha, 1994) should not lead us to forget that while we challenge the idea of bounded and fixed cultural groups, people are still cultural beings. Culture is still inescapably involved in our allocation to unequal positions in society, and culture is an important element of self-ascribed identity (Phillips, 2007), both of which impact on the experience of infertility and its treatment. The contributions to this book show how culture shapes the interaction between infertile people and ARTs, but also how this interaction is impacted by other factors.

The tension between wanting simultaneously to question the appropriateness of culturalist explanations and insist on the importance of culture for individuals is one that is reflected in the health inequalities literature more generally and to some extent is represented in this volume. It would not be correct to say that there is a strong, unified theoretical orientation in this collection. Chapters not only arise from different disciplines, but clearly employ somewhat varied understandings of the role of culture(s) and some have a distinctly more critical edge than others. What unites the authors, however, is a concern for exposing the hegemony of the dominant perspective of infertility as a concern of Western elites, through presenting the meanings of infertility from a range of diverse voices. This is coupled with a strong plea for providers of infertility services to carefully navigate the admittedly difficult terrain of taking into account the broader social and cultural context of the infertile patient, while avoiding stereotypical assumptions derived from essentialized ideas of cultural groups.

Structure of the book

The book is divided into two principal parts. In Part One, 'Researching Infertility, Ethnicity and Culture' the authors raise for debate, several important conceptual and methodological issues in research on infertility. In Part Two, 'Exploring Infertility, Ethnicity and Culture in National Contexts' the book examines the relationship between ethnicity, culture and reproductive technologies, and explores the impact of infertility on minority ethnic communities, through a discussion of primary research. It is not, of course, comprehensive. Nevertheless, it does reflect much of the important work currently underway in this emerging (though still marginal) field of enquiry.

The opening chapter of Part One surveys our current state of knowledge of ethnic differences in access to infertility services in Western societies, with particular emphasis on the UK and US. Wide disparities exist in the availability and quality of infertility services between the developed and less developed countries of the world. Even in wealthy countries access to ARTs is marked by inequality. Our understanding of this process, however, is limited. Evidence on ethnic differences in infertility prevalence in the UK and especially in Europe is sparse, reflecting a lack of attention in research studies to ethnic differentiation. Several European countries explicitly prohibit the collection of ethnicity data for political reasons (see Chapter 4). In the US, there has been more concern with 'racial' disparities in infertility care. However, despite a higher incidence of infertility among many minority groups in the US and strong cultural meanings attached to having children, the needs of minority women have received little attention academically or in public policy. The evidence reviewed by Lorraine Culley suggests that in the US there are multiple obstacles hindering minority ethnic couples from seeking medical care for infertility (structural, cultural, ideological and political), reflecting a wider process of 'stratified reproduction'. While evidence on access in the UK is sparse, the chapter explores the possible impact of ethnicity on access to ARTs and warns of the danger of proposing uniquely 'cultural' barriers to infertility treatment for members of minority ethnic groups.

The relevance of culture is further discussed in the second chapter, which explores variations in explanatory models of infertility, procreation beliefs and religious ideologies and discusses their potential impact on infertility. Frank van Balen discusses significant cultural differences in the consequences of childlessness and some of the possible implications of our values, attitudes and religious beliefs on seeking treatment for infertility. The chapter demonstrates the cultural variability of understandings of infertility and argues that treatment providers should have a greater awareness of differences in the meanings of and consequences of childlessness in different cultures.

The fact that healthcare providers fail to effectively engage with ethnic diversity is discussed more directly in Chapter 3. Karl Atkin takes us into the complex terrain of defining ethnic identity, the shifting nature of ethnicity (as it comes to embody language, religion, faith, culture, ancestry, nationality, histories of migration and a shared heritage), and the relationship between ethnicity and socio-economic status, age and gender. The ways in which ethnicity is also constructed by state practices and policies are also highlighted. With a specific focus on the UK, the author introduces the important concept of institutional racism, providing a framework within which we can understand the process whereby the needs and experiences of ethnic minority populations can become ignored or misrepresented.

The authors of Chapter 4 also consider the issue of institutional racism, this time in the form of an inadequate research base for tackling ethnic inequalities in health and healthcare. Mark Johnson and Theda Borde argue for the need to capture or record the 'ethnic' origins of people (both in specific research activity and in monitoring which permits and supports epidemiological research), and the need to obtain 'insider' accounts of health beliefs and behaviours of members of minority groups. Drawing on several research studies carried out in Germany and the UK, the authors describe some of the challenges in researching with minority ethnic populations, especially in a comparative context. Nevertheless, they provide a powerful argument for carrying out both detailed epidemiological and qualitative participatory research on ethnicity and infertility and offer the reader a guide to some useful questions to ask of any study of health and healthcare delivery.

The final chapter in the first part of the book addresses an issue which faces all social researchers, though it is often only raised in relation to research with minority ethnic communities. What is the impact of our identity as researchers? In particular, what difference does our difference make in social research? Yasmin Gunaratnam explores how such difference has been approached epistemologically, methodologically and ethically. Drawing on a range of literature and giving particular examples from infertility research, she explores the constraints constructed by the dualism inherent in the notions of 'insider' and 'outsider' and questions the rationale for ethnic 'matching' in the research process. She argues, nevertheless, that difference matters deeply in research and that attentiveness to difference can enhance our conceptual frameworks, methodological approaches and ethical standards.

Part Two begins with a discussion of the ways in which the infertility experience is shaped and lived out within British South Asian communities. Lorraine Culley and Nicky Hudson explore how communally constructed understandings of infertility are expressed, and how these constructions impact on those who experience infertility in this context. While recognizing the fact that we are all 'cultural beings' they argue against an essentialism that tends to ignore internal differentiation and deny agency. They argue for the need to recognize the importance of social class, gender and generation in any discussion of cultural differences in the experience of infertility. They also discuss the ways in which institutional discrimination impacts on the provision of infertility care to people of Pakistani, Bangladeshi and Indian origin in England.

This is followed by a discussion drawn from an ethnographic study of London's Turkish community by Zeynep Gürtin-Broadbent. She describes the importance of having children as a duty and an expectation, strongly endorsed by Islam. Within this family-based, tight-knit community, it is expected that children will follow soon after marriage and so childlessness in married couples is highly visible. The treatment seeking trajectories of Turkish couples are described, with a mixture of biomedical and religious 'solutions' commonly followed. The chapter shows the significance of gender and of religion in seeking and accepting infertility treatment, demonstrating both the importance of Islam and the individual moral manoeuvring which is characteristic of the lived practice of religion. The author also discusses the potential misunderstandings or miscommunications that can arise when doctors and patients are acting within different cultural frameworks and which, in this study, resulted in several participants seeking out Turkish doctors, both in the UK and in Turkey.

The experience of Turkish migrants in Europe is also discussed in the chapter by Floor van Rooij and Dineke Korfker. The authors report on two studies carried out in the Netherlands; one with both Turkish and Moroccan migrants and one with Turkish couples. The chapter focuses on the experiences of both communities in accessing infertility treatment, describing a failure on the part of medical practitioners to fully explain the nature of the treatment process and a failure to provide appropriate language support. As with the Turkish community in London, discussed in Chapter 7, sometimes this led to feelings of discrimination among Moroccan and Turkish infertility patients, some of whom also reported a lack of respect from healthcare providers. Also with strong similarities to the London Turkish community, Islam was a major influence on both attitudes to childlessness and approaches to treatment for Moroccan and Turkish migrants. Links with Turkey were also strong among the participants and many couples travelled to Turkey for advice and/or treatment, stressing the value of culturally and linguistically congruent care. This was less so for the Moroccan community, perhaps because Morocco has a less well developed ART sector.

The work of van Rooij and colleagues with Turkish migrants in the Netherlands (van Rooij et al, 2006) has also included an analysis of the cultural interpretations of fertility and childlessness, and this is the issue taken up by Pranee Liamputtong in Chapter 9. The chapter is based on ethnographic work among Hmong women and traditional healers in Melbourne, Australia and gives a detailed account of the continued significance of specific religious beliefs and patrilineal and patrilocal culture in which Hmong women only achieve respect and status within the family and clan when they produce children. The author also describes a strong preference for male children; sons provide for elderly parents and upon their death must honour and feed the deceased parents' spirits. Without this their souls cannot reach the ancestral pool in the other world which enables them to reincarnate. Women, however, are not allowed to perform ancestral worship and so if a woman cannot bear sons she is obliged to let her husband take another wife. Drawing on Marcia Inhorn's (1994) discussion of 'levels of causal proximity', the author discusses Hmong women's understandings of infertility and describes their attempts to seek help from a range of ethnomedical specialists such as medicine women and shamans. Although acknowledging a process of acculturation that has modified the expression of some cultural practices, the author argues that traditional beliefs continue to have an impact on the way in which Hmong families deal with infertility in the Australian context.

Chapter 10 discusses a socio-cultural context that is literally and metaphorically a world apart from the Hmong in Victoria, Australia. Noëlle Cotter draws on her longitudinal work with infertile couples in Ireland, and an analysis of survey data and public documents, to explore perceptions of the embryo in the Irish state. The author describes a country in which the influence of Catholic mores is still very evident, and discusses the implications of the constitutional protection of the 'unborn' for ARTs and the creation, storage and disposal of embryos in particular. The chapter discusses the construction of the embryo as a 'boundary object', identifying competing Irish discourses emanating from 'expert', 'religious', 'lay' and 'patient' views. Legal and scientific 'experts' are concerned about the regulatory vacuum surrounding ARTs. Religious authorities strongly reiterate the constitutional protection of the embryo from the moment of fertilization, wherever that takes place. Lay opinion, though more complex, would appear largely to construct the embryo as a moral entity and a potential citizen attracting constitutional protection. Perhaps the most fascinating aspect of this work, however, is the 'patient' perspective. Here we see for the first time the complex imaginings of the embryo among Irish people undergoing fertility treatment, demonstrating how 'whiteness' and 'culture' are not mutually exclusive.

The final chapter of the book returns us to a key theme of this collection. Three leading scholars of infertility, Marcia Inhorn, Rosario Ceballo and Robert Nachtigall, draw our attention once more to the fact that medicine and social science have helped to perpetuate a dominant narrative of infertility as a concern of wealthy, well-educated, white couples. Following a theorization of the situation of ethnic minorities as a case of 'stratified reproduction' par excellence, the authors present the findings of their own studies of infertility among three marginalized minority ethnic populations: African Americans; Latino/a Americans; and Arab Americans. These populations, which share a poor health status with many white working class Americans, face higher levels of reproductive impairment, yet are simultaneously 'despised as reproducers in a racist/classist/xenophobic society'. The authors describe, for all three populations, the overwhelming impact of infertility and the many obstacles that impede the ability of people to receive medical treatment: poverty; lack of insurance cover; low awareness of treatment services; and communication barriers. The chapter discusses the demoralizing and devastating impact of this situation on the lives of their participants. It ends with a plea, which all contributors to this volume would echo, for the research community to rectify the gross absence of work among minority ethnic populations and to assist in the struggle to combat this form of stratified reproduction.

Conclusion

Research rarely responds to the multi-ethnic nature of developed societies, and policy and healthcare practice struggle to engage with minority ethnic populations (Atkin et al, 2006). Despite greater levels of infertility and strong cultural meanings attached to having children, the need for infertility services among minority ethnic women and men has received very little attention politically or academically. In drawing attention to the needs and healthcare experiences of infertile people from a range of minority ethnic communities, we have tried to initiate a corrective to the dominance of narratives of hegemonic groups in infertility research in Western societies.

Several commentators have argued that the idea of fertility treatment in less developed countries of the world often evokes feelings of disbelief and discomfort, since the dominant image of such societies is that of 'over-population'. There are those who suggest that similar feelings pervade public perceptions of marginalized communities in Western societies and that a desire to limit the reproductive capacity of such groups has been more apparent than any effort to assist procreative choice. This is perhaps most obviously the case in the US, where several studies have shown that those seeking infertility care are predominantly college educated, relatively wealthy and white. The situation is more complex in countries with more equitable healthcare systems, though we know little about ethnic differences in infertility prevalence and access to treatment in most countries. Such studies as we have recourse to, many of which are represented in this volume, demonstrate the ethnocentricity of fertility services. They show that even when people from minority ethnic communities successfully access infertility treatment, they nevertheless may receive less than optimal care.

There is a clear need for well resourced public health programmes directed to the prevention of a condition which, for many people of all ethnicities, is preventable. We must avoid the danger of allowing a preoccupation with the invasive and expensive 'solutions' offered by many ARTs to detract attention from the need for concerted efforts to reduce ethnic inequalities in health status generally and for public policies that attack the fundamental, avoidable causes of infertility. At the same time, however, we argue that more equitable access to ARTs for minoritized and racialized groups, an insistence on respect for cultural difference, an understanding of religious concerns, the provision of appropriate communication support and culturally competent counselling are essential components of a transformatory politics of infertility. Our hope is that this book will encourage further research and constructive debate on these important issues.

Notes

1   This figure is based on prevalence of current infertility of 12 months' duration and is lower than typically cited. This study also showed remarkably similar rates between more and less developed (UN categorization) countries, although there are some striking differences between individual countries. The mechanisms contributing to infertility vary between countries. Most cases of infertility in Africa, for example, are due to infections (from sexually transmitted diseases, unsafe abortions and poor postpartum care). In the West there are fewer cases related to infection but there is a steady increase in age-related infertility that is not found in low-resource countries.

2   The depiction of the infertile as in danger of long-term psychic damage is widespread. See Sandelowski and de Lacey (2002) for a critique of this and other stereotypical representations.

3   Strictly speaking most couples are sub-fertile, since most have at least some chance of conceiving with medical assistance. However 'infertility' is the most widely used term in the medical and social science literature. The most common (medical) definition of infertility is failure to conceive after one year of unprotected intercourse. Infertility is commonly classified as 'primary' if a couple is unable to conceive, and 'secondary' if the woman is unable to maintain a pregnancy, or is unable to conceive again after having had a child.

4   The number of cycles per million inhabitants was highest in Israel (3260 cycles per million), followed by Denmark (2040), then by Northern Europe, France and Germany (> 1000). In most of the Latin American countries less than 100 cycles per million people were performed. The percentage of ART births was highest in Denmark and the Netherlands (more than 4 per cent) and lowest in Latin America (less than 0.1 per cent) (ESRHE, 2006).

5   Infertility is big business. Revenues from fertility treatment in the US were nearly $3 billion in 2002, not including the lawyers, equipment suppliers, counsellors and other 'ancillary' services (Spar, 2006). Expensive IVF and ICSI have become the treatment of first resort for many who have the ability to pay, despite the evidence that many infertility problems can be quickly and cheaply diagnosed and effectively treated with less invasive methods (see Ombelet et al, 2008).

6   We recognize the problematic nature of the binary implied by the terms Western/non-Western. We are not suggesting that these represent homogeneous and distinct entities, but highly differentiated, permeable, complex and inter-related spheres (Grewel and Kaplan, 2006). We also recognize the difficulties of counterposing ethnic minority and majority communities. In problematizing the position of minority groups in the politics of infertility, we are precisely trying to expose the differentiated nature of 'communities'. In discussing the relevance of culture in the experience of infertility we are arguing for a more nuanced concept of culture(s) and do not see ethnic minorities (or majorities) as bounded, undifferentiated and fixed cultural groups (Eisenberg and Spinner-Halev, 2005).

7   The use of the bodies of poor women as donors and surrogates for women of the 'North' demonstrates, it is argued, the intersection of patriarchal traditions, the scientific medical establishment and the global market (Steinberg, 1997; Mukherjee and Nadimipally, 2006). ARTs have been discussed as an example of the metaphorical and literal fragmentation and commodification of the body and its parts (Radin, 1996; Sharp, 2000) exemplified in the process of 'reproductive tourism'.

8   For a discussion of ARTs as a cultural practice that promotes race-based hierarchies see Quiroga (2007).

References

Ahmad, W. I. U. (1996) 'The trouble with culture', in D. Kelleher and S. M. Hillier (eds) *Researching Cultural Differences in Health*, Routledge, London

Akker, O. van den (2002) *The Complete Guide to Infertility: Diagnosis, Treatment, Options*, Free Association Books, London

Allan, H. (2007) 'Experiences of infertility: Liminality and the role of the fertility clinic', *Nursing Inquiry*, vol 14, no 2, pp132–139

Atkin, K., Rodney, A. and Cheater, F. (2006) 'Disability, chronic illness, fertility and minority ethnic young people: Making sense of identity, diversity and difference', in R. Balen and M. Crawshaw (eds) *Sexuality and Fertility Issues in Ill Health and Disability*, Jessica Kingsley, London

Becker, G. (2000) *The Elusive Embryo: How Women and Men Approach New Reproductive Technologies*, University of California Press, Berkeley, CA

Bhabha, H. (1994) *The Location of Culture*, Routledge, London

Boivin, J., Bunting, L., Collins, J. and Nygren, K. (2007) 'International estimates of infertility prevalence and treatment-seeking: Potential need and demand for infertility medical care', *Human Reproduction*, vol 22, no 6, pp1506–1512

Culley, L. (2006) 'Transcending transculturalism? Race, ethnicity and health-care', *Nursing Inquiry*, vol 13, no 2, pp144–153

Darling, M. Tyson (2006) 'Gender, new technologies and development', *Development*, vol 49, no 1, pp23–27

Dickenson, D. (2004) 'What feminism can teach global ethics', in R. Tong (ed) *Linking Visions: Feminist Bioethics, Human Rights and the Developing World*, Rowman and Littlefield, Lanham, MD

Eisenberg, A. and Spinner-Halev, J. (eds) (2005) *Minorities within Minorities: Equality, Rights, Diversity*, Cambridge University Press, Cambridge

ESRHE (2006) 'World report on ART (assisted reproductive technology) fact sheet', accessed online at www.eshre.com/emc.asp?pageId=807 on 3 October 2008

Fassin, D. (2001) 'Culturalism as ideology', in C. M. Obermeyer (ed) *Cultural Perspectives on Reproductive Health*, Oxford University Press, Oxford

Franklin, S. (1997) *Embodied Progress: A Cultural Account of Assisted Conception*, Routledge, London

Ginsberg, F. D. and Rapp, R. (1995) *Conceiving the New World Order: The Global Politics of Reproduction*, University of California Press, Berkeley, CA

Greil, A. L. (1991) *Not Yet Pregnant: Infertile Couples in Contemporary America*, Rutgers University Press, New Brunswick, NJ

Greil, A. L. (1997) 'Infertility and psychological distress: A critical review of the literature', *Social Science & Medicine*, vol 45, no 11, pp1679–1704

Grewel, I. and Kaplan, C. (2006) 'Introduction: Transnational feminist practices and questions of postmodernity', in I. Grewal and C. Kaplan (eds) *Scattered Hegemonies*, University of Minnesota Press, Minneapolis, MN

Hall, S. (1992) 'New ethnicities', in J. Donald and A. Rattansi (eds) *Race, Culture and Difference*, Sage, London

Hartouni, V. (1994) 'Breached birth: Reflections of race, gender, and reproductive discourse in the 1980s', *Configurations*, vol 2, no 1, pp73–88

Inhorn, M. C. (1994) *Quest for Conception: Gender, Infertility, and Egyptian Medical Tradition*, University of Pennsylvania Press, Philadelphia, PA

Inhorn, M. C. (1996) *Infertility and Patriarchy: The Cultural Politics of Gender and Family Life in Egypt*, University of Pennsylvania Press, Philadelphia, PA

Inhorn, M. C. (2003) *Local Babies, Global Science: Gender, Religion and In Vitro Fertilization in Egypt*, Routledge, New York

Inhorn, M. and van Balen, F. (eds) (2002) *Infertility around the Globe: New Thinking on Childlessness, Gender and Reproductive Technologies*, University of California Press, Berkeley, CA

Jenkins, G. L. and Inhorn, M. C. (2003) 'Reproduction gone awry: Medical anthropological perspectives', *Social Science and Medicine*, vol 56, no 9, pp1831–1836

Kelleher, D. (1996) 'A defence of the use of the terms of "ethnicity" and "culture"', in D. Kelleher and S. M. Hillier (eds) *Researching Cultural Differences in Health*, Routledge, London

Letherby, G. (1999) 'Other than mother and mothers as others: The experience of motherhood and non-motherhood in relation to "infertility" and "involuntary childlessness"', *Women's Studies International Forum*, vol 22, no 3, pp359–372

Monach, J. H. (1993) *Childless, No Choice: The Experience of Involuntary Childlessness*, Routledge, London

Mukherjee, M. and Nadimipally, S. (2006) 'Assisted reproductive technologies in India', *Development*, vol 49, no 4, pp128–134

Mulay, S. and Gibson, E. (2006) 'Marketing of assisted human reproduction and the Indian state', *Development*, vol 49, no 4, pp84–93

Nachtigall, R. D. (2006) 'International disparities in access to infertility services', *Fertility and Sterility*, vol 85, no 4, pp871–875

Nsiah-Jefferson, L. and Hall, E. J. (1989) 'Reproductive technology: Perspectives and implications for low-income women and women of colour', in K. S. Ratcliff (ed) *Healing Technology: Feminist Perspectives*, University of Michigan Press, Ann Arbor, MI, pp98–117

Ombelet, W., Cooke, I., Dyer, S., Serour, G. and Devroey, P. (2008) 'Infertility and the provision of infertility medical services in developing countries', *Human Reproduction Update*, pp1–17

Phillips, A. (2007) *Multiculturalism Without Culture*, Princeton University Press, Princeton, NJ

Phoenix, A., Woollett, A. and Lloyd, E. (1991) *Motherhood: Meanings, Practices and Ideologies*, Sage Publications, London

Quiroga, S. S. (2007) 'Blood is thicker than water: Policing donor insemination and the reproduction of whiteness', *Hypatia*, vol 22, no 2, pp143–161

Radin, M. (1996) *Contested Commodities: The Trouble with Trade in Sex, Children, Body Parts and Other Things*, Harvard University Press, Cambridge, MA

Reissman, C. K. (2000) 'Stigma and everyday practices: Childless women in south India', *Gender & Society*, vol 14, no 1, pp111–135

Roberts, D. (1997) *Killing the Black Body: Race, Reproduction, and the Meaning of Liberty*, Pantheon Books, New York

Rooij, F. van, van Balen, F. and Hermmans, J. (2006) 'Migrants and the meaning of parenthood: Involuntary childless Turkish migrants in the Netherlands,' *Human Reproduction*, vol 21, no 7, pp1832–1838

Sandelowski, M. (1993) *With Child in Mind: Studies of the Personal Encounter with Infertility*, University of Pennsylvania Press, Philadelphia, PA

Sandelowski, M. and de Lacey, S. (2002) 'The uses of "disease": Infertility as a rhetorical vehicle', in M. Inhorn and F. van Balen (eds) *Infertility around the Globe: New Thinking on Childlessness, Gender and Reproductive Technologies*, University of California Press, Berkeley, CA

Sharp, L. (2000) 'The commodification of the body and its parts', *Annual Review of Anthropology*, vol 29, pp287–328

Spar, D. L. (2006) *The Baby Business: How Money, Science, and Politics Drive the Commerce of Conception*, Harvard Business School Press, Boston, MA

Strathern, M. (1992) *Reproducing the Future: Essays on Anthropology, Kinship and the New Reproductive Technologies*, Manchester University Press, Manchester

Steinberg, D. (1997) *Bodies in Glass*, Manchester University Press, Manchester

Thompson, C. (2005) *Making Parents: The Ontological Choreography of Reproductive Technologies*, MIT Press, Cambridge, MA

Thompson, C. M. (2002) 'Fertile ground: Feminists theorize infertility', in M. Inhorn and F. van Balen (eds) *Infertility around the Globe: New Thinking on Childlessness, Gender, and Reproductive Technologies*, University of California Press, Berkeley, CA

Throsby, K. (2004) *When IVF Fails: Feminism, Infertility and the Negotiation of Normality*, Palgrave Macmillan, Basingstoke

Whiteford, L. M. and Gonzalez, L. (1995) 'Stigma: The hidden burden of infertility', *Social Science & Medicine*, vol 40, no 1, pp27–36

Widdows, H., Dickenson, D. and Hellsten, S. (2003) 'Global Bioethics', *New Review in Bioethics*, vol 1, no 1, pp101–116

Part One

Researching Infertility, Ethnicity and Culture

1

Dominant Narratives and Excluded Voices: Research on Ethnic Differences in Access to Assisted Conception in More Developed Societies

Lorraine Culley

Introduction

There is a considerable body of work in the UK, US, Australia, New Zealand and increasingly within Europe which documents inequities in the provision of healthcare to both migrant groups and settled immigrant or minority ethnic communities (Bhopal, 2007). Within the UK, the main emphasis in this field has been on access to primary care, coronary heart disease, diabetes, cancer and mental health services (Nazroo,1997; Smaje and Le Grand, 1997; Atkinson et al, 2001; Atkin, 2004).[1] In the area of reproductive health, considerably more emphasis has been placed on ethnic differences in access to sexual health and contraceptive services than infertility treatment (Katbamna, 2000), reflecting perhaps a dominant (racist) view that minority ethnic women are 'hyper-fertile'(Ceballo, 1999). In the US, Nsiah-Jefferson and Hall (1989) and Roberts (1999) have argued that ignoring the infertility problems of 'women of color' (and poor women) is consistent with the historical association between eugenics and social policy programmes that have promoted policies to maintain the dominance of the native-born white population, while restricting the reproductive activities of immigrants, blacks and native Americans.

In what follows I offer a commentary on our current broad state of knowledge of ethnic differences in access to infertility services in more developed societies. The review focuses on key studies from the US and the UK. It is not primarily concerned with ethnic differences in treatment *outcomes*. Suffice to say that research on ethnic differences in treatment outcomes is limited (especially in the UK) and findings are somewhat ambiguous (see Mahmud et al, 1995; Lashen et al, 1999; Sharara and McClamrock, 2000; Grainger et al, 2004; James et al, 2004; Bendikson et al, 2005; Feinberg et al, 2006; Purcell et al, 2007).[2] The chapter demonstrates that despite evidence for a higher incidence of infertility in minoritized groups (in the US at least) our knowledge of 'racial' and ethnic disparities in infertility treatment in the West is sparse and reflects both an absence of effective ethnic monitoring (especially in the UK) and a privileging of the experience of white couples in the social scientific literature on access to and experiences of assisted conception.

Ethnicity and infertility prevalence

Before commenting on issues of access to infertility services, it is useful to briefly review what we know about potential need, by exploring ethnic differences in infertility prevalence. A recent international review of population-based prevalence studies demonstrated a 12-month prevalence rate ranging from 3.5 to 16.7 per cent in more developed nations and from 6.9 to 9.3 per cent in less developed nations, with an estimated overall median prevalence of 9 per cent (Boivin et al, 2007). This study also confirms the fact that only about half of those who experience infertility problems seek medical care and less than a quarter actually receive treatment.

There have been relatively few studies in developed countries which examine the prevalence of infertility among different ethnic groups. In the US one early study reported a higher prevalence among 'Black' women (18 per cent) compared with white women (8 per cent) of the same age (Mosher and Aral, 1985). Data from a national survey of 10,847 women of reproductive age in the US found that Black, Hispanic and other 'non-Caucasian' women reported infertility more often than Caucasian women (10.5, 7, 13.6 and 6.4 per cent respectively) and that a significantly greater proportion of African-American and Hispanic women had tubal factor infertility compared with Caucasian women (Jain, 2006). Using pooled data for over 31,000 women from the US National Survey of Family Growth, Bitler and Schmidt (2006) also report higher rates of infertility for 'non-white' women than white women. African-American women are more likely to experience reproductive problems such as pelvic inflammatory disease and ectopic pregnancies, which are associated with infertility (Molock, 1999).

Most population-based prevalence studies in the UK include age as a variable, some studies include marital status, educational level, income level or other measure of socio-economic status, but few report ethnic origin. Studies in the countries of origin of migrant communities in the UK show raised levels of infertility, especially in sub-Saharan Africa and also in Asia (Zargar et al, 1997; Larsen, 2000). However, generalizations about settled diasporic populations based on rates in countries of ancestral origin are highly questionable. Whether we are exploring prevalence or service utilization, there are significant difficulties in interpreting data across different countries with variable healthcare systems and different approaches to ethnic classification.

Given the evidence of ethnic disparities in infertility prevalence in the US and our knowledge of the existence of ethnic inequalities in other health conditions in the UK (Aspinall and Jacobson, 2004), the research lacunae on prevalence among minority ethnic communities in the UK is remarkable. In the absence of published data, one can only comment on potentially significant contextual factors that might impact on ethnic patterning of infertility prevalence. First, fertility declines with age; the optimum age of fecundity is about 24 years for women, with a decline seen at age 25–30 and a more serious drop after the age of 30. In many developed societies, there is a growing incidence of delayed child-bearing, often associated with high levels of education and a high family income (van Balen et al, 1997). Although patterns are changing, South Asian women in the UK tend to marry at a younger age than the population as a whole and there is some evidence of earlier childbearing in most minority populations (Beishon et al, 1998). Age-related primary infertility may therefore, be reduced in these groups.

Second, a significant (preventable) cause of infertility is tubal damage follow-ing pelvic infection associated with sexually transmitted disease (STD). Several studies have shown a relatively low prevalence of reported STDs among Britain's South Asian communities which may be associated with a number of protective behaviours including delay in first sexual intercourse and reduction in life-time numbers of partners. Higher rates of STDs are found in people of African and African-Caribbean origin (Elam et al, 1999; Erens et al, 2001; Fenton and Wellings, 2001) and this is likely to have an impact on infertility rates.

Third, lifestyle factors (weight, smoking, diet, exercise, psychological stress, caffeine consumption, alcohol consumption and exposure to environmental pollutants) may have an impact on fertility (Chura and Norman, 2007; Homan et al, 2007) and these are likely to vary considerably between ethnic groups and within ethnic groups by socio-economic status. Studies have consistently demonstrated a low rate of smoking and drinking among South Asian women (Nazroo, 1997; Erens et al, 2001). The picture for males is more complex. While some South Asian males have lower rates of smoking, others such as those in the Bangladeshi community have much higher rates than the white population. Smoking rates are also high among Caribbean groups (Erens et al, 2001). The impact of smoking on male infertility, however, is uncertain (National Institute for Clinical Excellence (NICE), 2004). Women with a body mass index (BMI) of over 29 are likely to take longer to conceive (NICE, 2004). The evidence on ethnic differences in obesity shows a more complex picture than is often assumed. While obesity prevalence is high for Black Caribbean and Pakistani women, it is low for Bangladeshi and Chinese women. Indian women have a mean BMI similar to that of the general population (Erens et al, 2001). Rates for the Indian and Bangladeshi women aged 16–34 were lower than the population as a whole, while rates for the Pakistani group were slightly higher.

Fourth, two conditions that may impact disproportionately on the fertility of Asian women in particular are diabetes and Polycystic Ovary Syndrome (PCOS). Asian women are much more likely than white women to suffer from diabetes which may impact on fertility in severe cases (Bhopal, 2007). The prevalence of polycystic ovaries in South Asian women is high (Rodin et al, 1998).

Fifth, there is growing evidence of the impact of racism, both individual and institutional, on patterns of ill health among minority ethnic communities (Karlsen and Nazroo, 2002a; Nazroo, 2003; Karlsen and Nazroo, 2004; Dovidio et al, 2008). Racial harassment and discrimination, for example, can result in acute and chronic stress which produces physiological changes and problems for mental well-being. No research has been undertaken to explore the potential impact of this on the ability of women to conceive, but experience of racist verbal abuse or physical violence is related to a greater risk of premature death, high blood pressure, respiratory illness, lower self-esteem, psychological distress, depression and anxiety, stress and anger and psychosis (Karlsen, 2007). The effects of psychological stress on human reproductive function have been shown (Cwikel et al, 2004; Boivin et al, 2006) and it is not unreasonable, therefore, to suspect that stress arising from racism may be involved in infertility.

The overall impact of the above factors is difficult to assess. Some may lead to a lower prevalence of infertility in minority ethnic communities whilst others would suggest a higher prevalence. Although these factors may give some indication of possible influences on ethnic patterns in infertility, in the absence of proper epidemiological studies, it is impossible to make any systematic estimate of ethnic differences in the prevalence of infertility within UK populations. It is also important to note the diversity between different minority ethnic groups, both in health status and in relation to general socio-economic status (Nazroo, 1997). While there is a tendency, within the UK at least, to homogenize minority ethnic groups (using, for example the term 'black and minority ethnic' or BME) there are in fact significant differences between groups in both health status and experience of healthcare.[3]

Ethnicity and utilization of infertility treatment

When we examine the literature on the utilization of infertility services, we find very few studies that have explored ethnic differences. Only one study that has examined ethnicity in any detail has been published in the UK and this is discussed in detail below. Most work on access to assisted conception has been carried out in the US and this concentrates primarily on differences relating to socio-economic status and insurance coverage with relatively little consideration of ethnic differences (King and Meyer, 1997).[4]

Assisted conception and 'race' in the US

Where ethnicity or 'race' (the more common term in the US) has been included as a variable, American studies show a higher level of utilization of assisted conception by 'Caucasian' women than African-American and Hispanic women (Jain and Hornstein, 2005; Jain, 2006; Feinberg et al, 2007). Some authors maintain that there are 'cultural' barriers to accessing assisted conception for 'racial' minorities. White et al (2006), argue that in addition to structural barriers such as income:

> *African-American women's lower use of infertility treatment is probably attributable to factors such as cultural aversion to technological solutions to a holistic health problem, greater aversion to infertility labels and treatments, greater distrust of a medical establishment that has victimised minorities in the past, bad experiences with their own doctors, or a belief that they will be rejected for treatment.* (p856)

Jenkins (2005) also discusses cultural and religious 'barriers' to treatment in the US, but this paper does little more than highlight *potential* problems, since, as the author points out, there is very little research on attitudes to infertility and assisted conception in African Americans, Hispanic Americans and Asian Americans, and research on Native Americans is 'virtually non-existent' (p102).

In contrast, several studies suggest that differential uptake of assisted conception is primarily an effect of differences in socio-economic status and insurance cover. Stephen and Chandra (2000), for example, indicate that 'racial' differences in utilization of infertility services disappear once the effects of women's marital status, income and private health insurance coverage are taken into account. Jain (2006) found that significantly more Chinese women were seeking infertility treatment compared with their percentage in the general population and suggests that this reflects a higher level of education and income than other minority groups. Even in states where insurance coverage for infertility services is mandated there is still an under-representation of African-American and Hispanic women (Bitler and Schmidt, 2006; Jain, 2006; Feinberg et al, 2007). However, as Green et al (2001) show, African-American patients are much less likely than white patients to have insurance cover.

In the US, ART remains a private, fee-for-service form of healthcare delivery, inaccessible to low-income individuals, a group that includes a disproportionate number of 'people of color'. At the same time, there may be restrictions other than income, which limit treatment for some minority women, and it is important to explore the wider dimensions of 'unequal' utilization of assisted conception services. African-American women, for example, are much less likely to be married than white women and there are clinics that restrict treatment to married, heterosexual couples (Gurmankin et al, 2005).

Inhorn and Fakih (2006) take this debate further, suggesting that in the US an ideology of 'stratified reproduction' and a 'eugenic logic' results in ART being used to enhance the fertility of better-off white couples, while infertility is seen as a 'non-issue' for low-income and minority couples, who are stereotypically seen as being 'hyperfertile'. Government healthcare subsidies focus on reducing the fertility of minority and low-income people rather than improving it (Inhorn and Fakih, 2006). As Ceballo (2005) and Inhorn et al (Chapter 11) argue, public images of infertility are exclusionary – they hardly ever include 'women of color'. The hegemonic view of infertility, therefore, is of a trauma solely afflicting affluent whites.

Jain and Hornstein (2005) suggest the following potential barriers to access to fertility care in the US: lack of appropriate information; racial discrimination; lack of referrals from primary care; lack of insurance coverage; and cultural bias against treatment. However, as they point out, further studies are needed to explore the impact of these factors. It would appear that, as with access to medical care more generally in the US (Institute of Medicine, 2003; Williams et al, 2003), there are multiple barriers to accessing infertility treatment for 'racial minorities'.[5]

Dovidio et al (2008) suggest that black people generally have less trust in the healthcare system and healthcare providers than whites do, which may inhibit care seeking. Roberts (1999) also points out that black women may have historically based fears of technological intervention with reproduction, given that they have been the objects of sterilization abuse and medical experimentation and that black patients are deliberately steered away from ARTs by health professionals. She concludes that disparities in access may stem from 'a complex interplay of financial barriers, cultural preferences, and more deliberate professional manipulation' (p253).

Qualitative studies of ethnic inequalities in access to assisted conception in the US are rare. As Inhorn and Fakih (2006) point out, this is paradoxical given the possibility of a greater prevalence of infertility within some minority communities and the intense stigma which is attached to childlessness in the societies of origin of many migrants to the 'West' (Inhorn and van Balen, 2002). Ceballo (1999, 2005) has documented a hegemonic view of infertility as a trauma solely affecting white women, and a qualitative study of barriers to infertility care among African-American and Arab-American men in Michigan revealed significant barriers to effective infertility care, including economic constraints, cultural and linguistic barriers and social marginalization in US society, post-11 September 2001 (Inhorn and Fakih, 2006; Inhorn et al, Chapter 11). Becker et al (2006) report a rare ethnographic interview study of 118 'Latinos' who sought medical treatment for infertility in the San Francisco Bay Area. All the participants had limited economic resources and the clinic they accessed provided free medical care on one day a week, offering basic testing, diagnosis and 'low-tech' treatments only. No IVF or intrauterine insemination (IUI) was available. Only one couple had undergone IVF treatment elsewhere and very few couples had the resources to use private medical treatment once they had exhausted the limited treatments available. This study concludes that the strong cultural expectation of parenthood means that infertility is a devastating experience for Latinos, with far-reaching effects on men and women and on the couple relationship. These authors also make the important point that given the high levels of distress created by the stigma of infertility it is likely that those disadvantaged couples from minority ethnic communities who do not seek infertility care refrain from doing so primarily because of a lack of awareness of services and/or a lack of economic resources (Becker et al, 2006).

Infertility, ethnicity and access to treatment in the UK

In the US, the pattern of inequality in the utilization of infertility services mirrors the pervasive disparities in access to healthcare more generally (Williams et al, 2003). In the UK, with a publicly funded system of healthcare, free at the point of use, one might expect a different picture. However, the situation with regard to assisted conception is complicated by the fact that while infertility investigations and some treatments are available within the NHS, the majority of IVF treatment (75 per cent) is paid for by patients, and is costly. NHS-funded IVF falls far short of demand and very strict eligibility criteria apply. Despite the existence of national guidelines on the management of infertility, NHS treatment is locally commissioned and its availability varies from one locality to another (commonly referred to as a 'postcode lottery', see BioNews, 2007). The failure to publicly fund IVF is likely, then, to reduce access to treatment for economically disadvantaged groups. Some ethnic minority communities are disproportionately represented in such groups, and are thus more likely to face economic barriers to treatment. Over 60 per cent of households in Pakistani and Bangladeshi communities, for example, live below the poverty line (Platt, 2002).

Even within NHS-funded provision, however, the non-medical restrictions on eligibility for treatment may work disproportionately against minority ethnic groups. In our study of British South Asian couples undergoing assisted conception, a significant proportion had secondary infertility (Culley et al, 2004). NHS criteria rule out treatment for anyone who already has a child, even where this is a child from a previous relationship. Within some South Asian communities, there is a higher fertility rate, and having one child (especially if this is a female) may still be regarded as contravening strong social norms (Culley and Hudson, 2006). There is also a severe shortage of gametes from South Asian donors (Golombok and Murray, 1999; Human Fertilisation and Embryology Authority (HFEA), 2004) and a reluctance on the part of some clinics to consider offering gametes from 'unmatched' donors (Culley et al, 2004). The extent of ethnic disparities in access to infertility treatment in the UK, however, is unknown, since there are no published data available on the ethnic background of infertility patients in the NHS or in independent clinics.

Equity and access

Utilization of healthcare, of course, is not the same thing as access. Access is a complex (and contested) concept. It implies a consideration of need, demand and supply. An equitable service is usually defined as one that provides equal access for equal need (Aspinall and Jacobson, 2004). The evidence on ethnic differences in 'need' for infertility services (considered as prevalence), as we have seen, is unclear. The evidence on supply of services certainly suggests restricted access to many forms of treatment on the NHS and often long waiting lists, although the 'supply' in the independent sector is abundant and instantly available for most treatments. Even where there is a 'need' and adequate 'supply', there may be constraints on 'demand'. Demand can be inhibited by an economic barrier, as we have seen. It may also be inhibited by a 'lack of awareness' of services. Rather than framing this in a knowledge deficit model, however, this could be conceptualized as an effect of the failure of services to provide appropriate information about treatment possibilities to potential consumers. The studies by Culley et al (2004, 2006a) and a review of 'public' perceptions of gamete donation by Hudson et al (2009) suggest a failure to engage minorities in debates relating to assisted conception.

While economic barriers are likely to be significant for many minority couples, there is also the possibility of cultural or religious beliefs effectively restricting demand for treatment, either because fertility treatment is regarded as unacceptable or because culturally appropriate alternatives (e.g. informal adoption, remarriage, acceptance, etc.) are promoted. At the same time, however, it could be that the high value placed on children, the importance of parenthood for achieving full adult status and the serious social consequences of childlessness in some minority communities (Culley et al, 2004) might lead to a heightened demand for fertility treatment and possibly treatment seeking at an early stage.

Such research questions are the domain of the social sciences. The scant *epidemiological* literature on ethnicity and infertility in the West, which we have outlined above, could reflect a lack of attention to racial/ethnic disparities more generally until relatively recently (Ahmad and Bradby, 2007; Bhopal, 2007). Perhaps, then, within the social sciences, with a longstanding interest in race/ethnicity, we find a better record of research? Sadly, this is not the case. Social scientific studies of assisted conception have typically failed to include participants from minority ethnic groups and there are few studies that focus specifically on the experience of minority ethnic couples undergoing fertility treatment.

One such study is that of Culley et al (2004). In this project we explored the provision of infertility services to British South Asian communities. The study included interviews with 50 infertile people (37 women and 13 men) from British South Asian communities (Indian, Pakistani and Bangladeshi ethnicity), and a focus group study of community perceptions of fertility, childlessness and attitudes to fertility treatment (Culley et al, 2006a, 2007; Culley and Hudson, Chapter 6). In approaching this issue, we took the view that lay understandings of infertility, why it arises and how it might be overcome, may have an impact on treatment seeking behaviour. Our research revealed several important features of South Asian community perceptions of infertility which may impact on demand. First, South Asian communities were highly pronatalist: children are highly desired; parenthood is culturally mandatory and childlessness socially unacceptable (Culley and Hudson, 2006), although the intensity of the pressure to reproduce was seen to vary somewhat by community, and by social class. Infertility is, therefore, a highly stigmatizing condition for men and especially for women.

Second, when asked about causes of infertility, participants described various reasons why infertility may occur, including religious, cultural, behavioural or biological causes. Most participants in this study were aware that infertility might be amenable to medical treatment and were of the opinion that (while there was a good deal of expressed dissatisfaction with GPs generally) most couples would go to their family doctor for help, despite potential embarrassment and concerns about confidentiality. This it seems, reflects the fact that while treatment may be stigmatized, childlessness is an even more socially damaging 'deviant' status.

Third, there was also a widespread knowledge of the existence of IVF (in common with the general population) and community support for using assisted conception, at least where the couples own gametes were involved. Although many non-Western 'treatments' and the importance of prayer were discussed, most people felt that these would be largely complementary to biomedical treatment (Culley et al, 2007). Religious and cultural dimensions of accessing treatment were also considered in our follow-up study of public perceptions of gamete donation in the same communities (Culley et al, 2006b; Culley and Hudson 2007; Culley and Hudson, forthcoming). Here we found great concern about the use of donated sperm in fertility treatment, although there was more ambivalence about the use of donated eggs. Muslim groups in particular demonstrated a specific concern for the religious acceptability of third party-assisted conception, *haram* (forbidden) in Sunni Islam (see Inhorn, 2006).

Exploring how infertility and its treatment are perceived more generally within a socio-cultural context, then, is important. Such perceptions will have many implications for infertile couples in terms of help-seeking behaviour and also the broader treatment experience. The social context may influence, for example, decisions about treatment options and about the disclosure of treatment (Culley et al, 2004, 2007). Cultural and religious factors therefore are clearly of relevance in accessing and experiencing infertility treatment. However, our study also shows the intersection of culture with socio-economic status, gender and generation and suggests that the relevance of cultural and religious issues for any specific individual or couple cannot be assumed from a formal identification with a particular Census category or statement of membership of a religious grouping (see Culley and Hudson, Chapter 6).

The specific significance of religion and spirituality for access to treatment in Western societies has not been widely investigated empirically. Most discussions of the impact of religion on accessing assisted conception are abstract discussions of the official views of various 'world religions' (Schenker, 2000; Roudsari et al, 2007). Empirical research demonstrating the significance of religion is primarily focused on non-Western societies, such as the important work of Marcia Inhorn, derived from fieldwork in the Middle East (Inhorn, 1994, 1996, 2003, 2006). Limited evidence on minorities in the West suggests that adherence to specific religious beliefs may impact on the infertility experience in a number of complex ways (Richards, 2003; van Rooij et al, 2004, 2006, 2007; Ahmed, 2005; Culley et al, 2006b). As argued above, while there are undoubtedly religious considerations for many couples in responding to infertility, the impact of these cannot be assumed in any specific case. Moreover, many of the religious constraints and difficulties that some groups face in the healthcare system are the result of the failure of the service to respond appropriately to religiously derived requirements, such as the provision of female doctors for devout Muslim women.

More generally, there is a danger in proposing uniquely 'cultural' barriers to treatment as operating for people from minority ethnic groups (in contrast with 'majority' groups), whose attitudes to health issues and help-seeking behaviour are seen as being determined by 'their' culture to the exclusion of a consideration of other explanations. In describing the experience of racialized minorities there is often, as Phillips (2007) has argued, too much explanatory weight allotted to 'culture' and a tendency to ignore human agency and autonomy.

Inasmuch as there are 'cultural' barriers to access, critical research suggests that we could frame this as a failing on the part of the NHS to provide appropriate services rather than as a consequence of inherent characteristics of alleged 'cultural groups' (Gerrish, 2001; Atkin, Chapter 3). Healthcare providers may offer a poorer service to minority ethnic clients for a variety of reasons. There may be individual racism on the part of prejudiced healthcare staff and/or there may be insufficient attention paid to the specific linguistic or cultural needs of some members of minority groups which could be regarded as indirect or institutional racism (Bradby, 2001; Atkin, 2004; Gerrish et al, 2004). Evidence for the existence of both can be found within the British NHS (Johnson, 2004; Karlsen, 2007). In our study of British South Asian communities and infertility services, we found little understanding of cultural difference among healthcare professionals and a serious failure on the part of service providers to ensure adequate communication support for non-English speakers (Culley et al, 2006a). The UK Department of Health's own patient surveys reveal higher levels of dissatisfaction with the NHS among some minority ethnic groups compared with the white majority (Raleigh et al, 2004).

Conclusion

Understanding differences in access to healthcare is highly complex. Services can be needed, but neither demanded nor supplied; they can be both needed and demanded but not supplied and they may be both needed and supplied, but not demanded (Smaje and Field, 1997). The role of research here is to systematically appraise this range of factors affecting people's use of services.

In the US, the evidence on prevalence suggests a higher rate of infertility among minority ethnic groups and the limited evidence on access to treatment suggests that there are many obstacles which hinder minority ethnic couples from seeking medical care for infertility, producing a process of 'stratified reproduction' (Inhorn et al, Chapter 11). In the UK, tackling inequality and promoting good practice in healthcare has been a key policy objective of successive governments (Department of Health, 2000, 2003). In this context, then, the paucity of research on ethnic differences in infertility prevalence, access to treatment and treatment outcomes is remarkable. This chapter has shown that in the UK in particular, studies of infertility prevalence rarely comment on ethnicity; studies of ethnicity and infertility treatment outcome are few and of questionable quality; studies of ethnic differences in access to assisted conception are virtually non-existent.[6]

As Bhopal (2007) has forcefully argued, health and healthcare data that can be analysed by ethnicity (in sufficiently detailed form, which may include not only ethnic origin but other relevant items such as religion and language) are essential to establish the extent of health inequalities, to define inequity in health service provision, monitor the impact of interventions to reduce inequalities, tackle racism and make good decisions based on evidence. Appropriate ethnicity data on infertility prevalence and access to assisted conception, coupled with qualitative studies of the experience of infertility and its treatment (Peddie and van Teijlingen, 2005) that include participants from minority ethnic communities, would greatly improve our understanding of the relationship between ethnicity and infertility and hopefully lead to improved healthcare practice.

Notes

1   The impact of access on health differentials is under-researched and there is a dearth of evidence of 'what works' in attempting to intervene to enhance equity of access (Aspinall and Jacobson, 2004).

2   Many studies of treatment outcome within the UK medical literature display a poor operationalization of 'ethnicity' variables, reflecting the inadequacies of biological reductionism within clinical studies (for a discussion see Aspinall, 2001, 2002; Ellison, 2005; Outram and Ellison, 2006; Moscou, 2008).

3   Within the literature on ethnic health inequalities, while the epidemiology of ethnic differentials in health status is well documented for several conditions, the *causes* of ethnic inequalities remain contested (Aspinall and Jacobson, 2004). Early accounts stressed the significance of cultural attributes in producing ethnic inequalities, in a clearly 'victim blaming' approach. More recently, both descriptive and analytical epidemiological evidence points to a significant contribution made by socio-economic differentials and racism, rather than genetic or cultural factors (Nazroo, 1997, 1998; Smith, 2000; Cooper, 2002; Karlsen and Nazroo, 2002a, 2002b).

4   Several important studies have explored other 'exclusions' such as those based on marital status and sexual orientation (Steinberg, 1997; Saffron, 2002; Hunfeld et al, 2004; Woodward and Norton, 2006) and HIV status (Pitts and Shields, 2004).

5   A small number of studies of the experience of minorities in infertility care have also been carried out in Europe, with particular emphasis on Turkish migrant communities in the Netherlands and Germany (Yüksel, 1995; Kentenich and Yüksel, 1997; Gacinski et al, 2002; van Rooij et al, 2004, 2006, 2007). The emphasis here has been on the younger age of treatment seeking, and the greater level of emotional distress of migrants compared with 'indigenous' populations, linked primarily to a strong social norm to have children, rather than on issues of access per se. Many Turkish migrant couples also return to Turkey to seek treatment, which may relate to cultural and linguistic congruity (van Rooij and Korfker, Chapter 8). Yebei (2000) explores the difficulties experienced by migrant Ghanaian women in the Netherlands, resulting from poor insurance coverage, low socio-economic status and language barriers. A large scale study in Sweden (where three cycles of IVF are provided by the national health insurance system) found that all groups of 'foreign born' women (Sweden does not collect ethnicity data) sought medical help at least as much as Swedish-born women (Eggert et al, 2008).

6   There is an extensive literature on the psychological consequences of infertility and infertility treatment in the West which is beyond the remit of this chapter. Few papers, however, reflect on the ways in which 'race' is deployed in biomedical solutions to infertility and most studies focus on the experiences of white, middle class treatment seekers (Greil, 1997).

References

Ahmad, W. I. U. and Bradby, H. (2007) 'Locating ethnicity and health: Exploring concepts and contexts', *Sociology of Health and Illness*, vol 29, no 6, pp795–810

Ahmed, M. (2005) 'A comparison of the psychosocial impact of infertility and its treatment on Pakistani and white couples in the UK', unpublished doctoral thesis, University of Leeds, Leeds

Aspinall, P. (2001) 'Operationalising the collection of ethnicity data in studies of the sociology of health and illness', *Sociology of Health and Illness*, vol 23, no 6, pp829–862

Aspinall, P. (2002) 'Collective terminology to describe the minority ethnic population: The persistence of confusion and ambiguity in usage', *Sociology*, vol 36, no 4, pp803–816

Aspinall, P. and Jacobson, B. (2004) *Ethnic Disparities in Health and Healthcare*, Department of Health, London

Atkin, K. (2004) 'Institutional racism, policy and practice', in S. Ali and K. Atkin (eds) *Primary Healthcare and South Asian Populations: Meeting the Challenges*, Radcliffe Medical Press Ltd, Abingdon

Atkinson, M., Clark, M., Clay, D., Johnson, M., Owen, D. and Szczepura, A. (2001) *Systematic Review of Ethnicity and Health Service Access for London*, University of Warwick, Warwick

Balen, F. van, Verdurmen, J. E. and Ketting, E. (1997) 'Age, the desire to have a child and cumulative pregnancy rate', *Human Reproduction*, vol 12, no 3, pp623–627

Becker, G., Castrillo, M., Jackson, R. and Nachtigall, R. D. (2006) 'Infertility among low-income Latinos', *Fertility and Sterility*, vol 85, pp882–887

Beishon, S., Modood, T. and Virdee, S. (1998) *Ethnic Minority Families*, Policy Studies Institute, London

Bendikson, K., Cramer, D. W., Vitonis, A. and Hornstein, M. D. (2005) 'Ethnic background and in vitro fertilization outcomes', *International Journal of Gynecology and Obstetrics*, vol 88, pp342–346

Bhopal, R. S. (2007) *Ethnicity, Race and Health in Multicultural Societies*, Oxford University Press, Oxford

BioNews (2007) 'Fertility treatment in UK remains inadequate', www.bionews.org.uk/ new.lasso?storyid=3539, accessed 18 April 2008

Bitler, M. and Schmidt, L. (2006) 'Health disparities and infertility: Impacts of state-level insurance mandates', *Fertility and Sterility*, vol 85, pp858–865

Boivin, J., Sanders, K. and Schmidt, L. (2006) 'Age and social position moderate the effect of stress on fertility', *Evolution and Human Behavior*, vol 27, no 5, pp345–356

Boivin, J., Bunting, L., Collins, J. A. and Nygren, K. G. (2007) 'International estimates of infertility prevalence and treatment-seeking: Potential need and demand for infertility medical care', *Human Reproduction*, vol 22, pp1506–1512

Bradby, H. (2001) 'Communication, interpretation and translation', in L. Culley and S. Dyson (eds) *Ethnicity and Nursing Practice*, Palgrave, Basingstoke

Ceballo, R. (1999) '"The only black woman walking the face of the earth who cannot have a baby": Two women's stories', in M. Romero and A. J. Stewart (eds) *Women's Untold Stories: Breaking Silence, Talking Back, Voicing Complexity*, Routledge, New York

Ceballo, R. (2005) 'Coping in silence: Experiences among African American women', paper presented at Reproductive Disruptions: Childlessness, Adoption and Other Reproductive Complexities, University of Michigan, Ann Arbor, MI

Chura, L. and Norman, R. J. (2007) 'Impact of lifestyle factors on ovarian function and reproductive health in women', *Women's Health*, vol 3, pp511–513

Cooper, H. (2002) 'Investigating socio-economic explanations for gender and ethnic inequalities in health', *Social Science and Medicine*, vol 54, no 5, pp693–706

Culley, L. and Hudson, N. (2006) 'Disrupted reproduction and deviant bodies: Pronatalism and British South Asian communities', *International Journal of Diversity in Organisations, Communities and Nations*, vol 5, pp117–126

Culley, L. and Hudson, N. (2007) 'Public understandings of science: British South Asian men's perceptions of third party assisted conception', *International Journal of Interdisciplinary Social Sciences*, vol 2, no 4, pp79–86

Culley, L. and Hudson, N. (2009) 'Constructing relatedness: Ethnicity, gender and third party assisted conception in the UK', *Current Sociology*, vol 57, no 2, Monograph 1, pp257–275

Culley, L., Rapport, F., Johnson, M., Katbamna, S. and Hudson, N. (2004) *Improving Policy and Practice: A Study of the Provision of Infertility Services to South Asian Communities*, Report to Dept of Health, De Montfort University, Leicester

Culley, L., Hudson, N., Rapport, F., Katbamna, S. and Johnson, M. (2006a) 'British South Asian communities and infertility services', *Human Fertility*, vol 9, pp37–45

Culley, L. Hudson, N., Johnson, M., Rapport, F. and Bhardadwaj, A. (2006b) *Public Perceptions of Gamete Donation in British South Asian Communities: Final Report to the ESRC*, De Montfort University, Leicester

Culley, L., Hudson, N., Johnson, M., Rapport, F. and Katbamna, S. (2007) '"I know about one treatment where they keep the egg somewhere": British South Asian community understandings of infertility and its treatment', *Diversity in Health and Social Care*, vol 4, pp113–121

Cwikel, J, Gidion, Y. and Sheiner, E. (2004) 'Psychological interactions with infertility among women', *European Journal of Obstetrics & Gynecology and Reproductive Biology*, vol 117, no 2, pp126–131

Department of Health (2000) *The Vital Connection: An Equalities Framework for the NHS*, Department of Health, London

Department of Health (2003) *Equalities and Diversity Strategy and Delivery Plan to Support the NHS*, Department of Health, London

Dovidio, J., Penner, L., Albrecht, T., Norton, W., Gaertner, S. and Shelton, J. (2008) 'Disparities and distrust: The implications of psychological processes for under-standing racial disparities in health and health care', *Social Science and Medicine*, vol 67, pp478–486

Eggert, J, Li, X. and Sundquist, K. (2008) 'Country of birth and hospitalization for pelvic inflammatory disease, ectopic pregnancy, endometriosis, and infertility: A nationwide study of 2 million women in Sweden', *Fertility and Sterility*, vol 90, no 4, pp1010–1025

Elam, G., Fenton, K., Nazroo, J., Johnson, A. and Ritchie, J. (1999) *Exploring Ethnicity and Sexual Health*, National Centre for Social Research, London

Ellison, G. (2005) '"Population profiling" and public health risk: When and how should we use race/ethnicity?', *Critical Public Health*, vol 15, no 1, pp65–74

Erens, B., Primatesta, P. and Prior, G. (2001) *Health Survey for England. The Health of Minority Ethnic Groups 1999, vol 1*, The Stationery Office, London

Feinberg, E. C., Larsen, F. W., Catherino, W. H., Zhang, Z. and Armstrong, A. Y. (2006) 'Comparison of assisted reproductive technology utilization and outcomes between Caucasian and African American patients in an equal-access-to-care setting', *Fertility and Sterility*, vol 85, pp888–894

Feinberg, E. C., Larsen, F. W., Wah, R. M., Alvero, R. and Armstrong, A. Y. (2007) 'Economics may not explain Hispanic underutilization of assisted reproductive tech-nology services', *Fertility and Sterility*, vol 88, no 5, pp1439–1441

Fenton, K. A. and Wellings, K. (2001) 'Sexual health and ethnicity', in H. Macbeth and P. Shetty (eds) *Health and Ethnicity*, Taylor and Francis, London

Gacinski, L., Yüksel, E. and Kentenich, H. (2002) 'The unfulfilled desire for a child of oriental couples. Infertility counselling and treatment for Turkish immigrants in Germany', in B. Strauss (ed) *Involuntary Childlessness, Psychological Assessment, Counseling and Psychotherapy*, Hogrefe & Huber, Seattle, WA

Gerrish, K. (2001) 'The nature and effect of communication difficulties arising from interactions between district nurses and South Asian patients and their carers', *Health and Nursing Policy Issues*, vol 33, no 5, pp566–574

Gerrish, K., Chau, R., Sobowale, A. and Birks, E. (2004) 'Bridging the language barrier: The use of interpreters in primary care nursing', *Health and Social Care in the Community*, vol 12, no 5, pp407–413

Golombok, S. and Murray, C. (1999) *Egg and Semen Donation: A Survey of UK Licensed Centres: Report for the National Gamete Donation Trust*, City University, London

Grainger, D. A., Seifer, D. B., Frazier, L. M., Rall, M. J., Tjaden, B. L. and Merrill, J. C. (2004) 'Racial disparity in clinical outcomes from women using advanced reproductive technologies (ART): Analysis of 80,196 ART cycles from the SART database 1999 and 2000', *Fertility and Sterility*, vol 82, suppl. 37–38

Green, J. A., Robins J. C., Scheiber, M., Awadalla, S. and Thomas, M. A. (2001) 'Racial and economic demographics of couples seeking infertility treatment', *American Journal of Obstetrics and Gynecology*, vol 184, pp1080–1082

Greil, A. L. (1997) 'Infertility and psychological distress: A critical review of the literature', *Social Science and Medicine*, vol 45, no 11, pp1679–1704

Gurmankin, A., Caplan, A. and Braverman, A. (2005) 'Screening practices and beliefs of assisted reproductive technology programs', *Fertility and Sterility*, vol 83, pp61–67

HFEA (2004) 'Sperm, Egg and Embryo Donation (SEED) review. Findings of the clinic survey', http://hfea.org.uk/aboutHFEA/HFEAPolicy/SEEDreview, accessed 15 September 2004

Homan, G. F., Davies, M. and Norman, R. (2007) 'The impact of lifestyle factors on reproductive performance in the general population and those undergoing infertility treatment: A review', *Human Reproduction*, vol 13, pp209–223

Hudson, N., Culley, L., Rapport, F., Johnson, M. and Bharadwaj, A. (2009) '"Public" perceptions of gamete donation: A research review', *Public Understanding of Science*, vol 18, no 1, pp61–77

Hunfeld, J., Passchier, J., Bolt, L. and Buijsen, M. (2004) 'Protect the child from being born: Arguments against IVF from heads of the 13 licensed Dutch fertility centres, ethical and legal perspectives', *Journal of Infant and Reproductive Psychology*, vol 22, no 4, pp279–289

Inhorn, M. (1994) *Quest for Conception. Gender Infertility and Egyptian Medical Traditions*, University of Pennsylvania Press, Philadelphia, PA

Inhorn, M. (1996) *Infertility and Patriarchy. The Cultural Politics of Gender and Family Life in Egypt*, University of Pennsylvania Press, Philadelphia, PA

Inhorn, M. (2003) *Local Babies, Global Science*, Routledge, London

Inhorn, M. (2006) 'Fatwas and ARTs: IVF and gamete donation in Sunni v. Shi'a Islam', *Journal of Gender, Race and Justice*, vol 9, pp291–317

Inhorn, M. C. and Fakih, M. H. (2006) 'Arab Americans, African Americans and infertility: Barriers to reproduction and medical care', *Fertility and Sterility*, vol 85, pp844–852

Inhorn, M. C. and van Balen, F. (eds) (2002) *Infertility around the Globe: New Thinking on Childlessness, Gender and Reproductive Technologies*, University of California Press, Berkeley, CA

Institute of Medicine (2003) *Unequal Treatment: Confronting Racial and Ethnic Disparities in Health Care*, National Academies Press, Washington, DC

Jain, T. (2006) 'Socioeconomic and racial disparities among infertility patients seeking care', *Fertility and Sterility*, vol 85, pp876–881

Jain, T. and Hornstein, M. D. (2005) 'Disparities in access to infertility services in a state with mandated insurance coverage', *Fertility and Sterility*, vol 84, pp221–223

James, C. E., Hammind, K. R. and Steinkampf, M. P. (2004) 'Race and assisted reproduction: A case-controlled study of outcomes in African-American and Caucasian women', *Fertility and Sterility*, vol 82, suppl. s123

Jenkins, R. L. (2005) 'Ensuring access to education and services on infertility for the underserved', *Journal of the National Cancer Institute Monographs*, vol 34, pp101–103

Johnson, M. (2004) *'Racial' and Ethnic Inequalities in Health. A Critical Review of the Evidence*, Centre for Evidence in Ethnicity, Health and Diversity, University of Warwick, Coventry and De Montfort University, Leicester

Karlsen, S. (2007) *Ethnic Inequalities in Health: The Impact of Racism*, Better Health Briefing 3, Race Equality Foundation, London

Karlsen, S. and Nazroo, J. (2002a) 'Agency and structure: The impact of identity and racism on the health of ethnic minority people', *Sociology of Health and Illness*, vol 24, no1, pp1–20

Karlsen, S. and Nazroo, J. (2002b) 'Relation between racial discrimination, social class and health among ethnic minority groups', *American Journal of Public Health*, vol 92, pp624–631

Karlsen, S. and Nazroo, J. (2004) 'Fear of racism and health', *Journal of Epidemiology and Community Health*, vol 58, no 12, pp1017–1018

Katbamna, S. (2000) *'Race' and Childbirth*, Open University Press, Buckingham

Kentenich, H. and Yüksel, E. (1997) *Psychosomatisches Betreuungskonzept Steriler Turkischer Paare in der Migration [Psychosomatic Concept of Care for Migrated Sterile Turkish Couples]*, Virchow-Klinikum, medizinische fakultat der HUB, Frauen- und Poliklinik / Frauen- und Kinderklinik, DRK-Klinikern Westend, Berlin

King, K. and Meyer, H. M. (1997) 'The politics of reproductive benefits. US insurance coverage of contraceptive and infertility treatments', *Gender and Society*, vol 11, pp8–30

Larsen, U. (2000) 'Primary and secondary infertility in sub-Saharan Africa', *International Journal of Epidemiology*, vol 29, pp285–291

Lashen, H., Afnan, M. and Sharif, K. (1999) 'A controlled comparison of ovarian response to controlled stimulation in first generation Asian women compared with white Caucasians undergoing *in vitro* fertilisation', *British Journal of Obstetrics and Gynaecology*, vol 106, pp407–409

Mahmud, G., Bernal, A. L., Yudkin, P., Ledger, W. and Barlow, D. (1995) 'A controlled assessment of the in vitro fertilization performance of British women of Indian origin compared with white women', *Fertility and Sterility*, vol 64, no 1, pp103–106

Molock, S. (1999) 'Racial, cultural and religious issues in infertility counselling', in L. Burns and A. Covington (eds) *Infertility Counseling: A Comprehensive Handbook for Clinicians*, Parthenon Publishing, New York

Moscou, S. (2008) 'The conceptualization and operationalization of race and ethnicity by health services researchers', *Nursing Inquiry*, vol 15, no 2, pp94–105

Mosher, W. D. and Aral, S. O. (1985) 'Factors related to infertility in the United States, 1965–1976', *Sexually Transmitted Diseases*, vol 12, no 3, pp117–123

National Institute for Clinical Excellence (NICE) (2004) *Fertility: Assessment and Treatment for People with Fertility Problems: Clinical Guideline 11*, National Institute for Clinical Excellence, London

Nazroo, J. Y. (1997) *The Health of Ethnic Minorities*, Policy Studies Institute, London

Nazroo, J. Y. (1998) 'Genetic, cultural or socio-economic vulnerability? Explaining ethnic inequalities in health', *Sociology of Health and Illness*, vol 20, no 5, pp710–730

Nazroo, J. Y. (2003) 'The structuring of ethnic inequalities in health: Economic position, racial discrimination, and racism', *American Journal of Public Health*, vol 93, pp277–284

Nsiah-Jefferson, L. and Hall, E. (1989) 'Reproductive technology: Perspectives and implications for low income women and women of color', in K. S. Ratcliff (ed) *Healing Technologies. Feminist Perspectives*, University of Michigan Press, Ann Arbor, MI

Outram, S. and Ellison, G. (2006) 'Anthropological insights into the use of race/ethnicity to explore genetic contributions to disparities in health', *Journal of Biosocial Science*, vol 38, pp83–102

Peddie, V. L. and van Teijlingen, E. (2005) 'Qualitative research in fertility and reproduction: Does it have any value?', *Human Fertility*, vol 8, pp263–267

Phillips, A. (2007) *Multiculturalism Without Culture*, Princeton University Press, Princeton, NJ

Pitts, M. and Shields, P. (2004) 'Access to infertility investigations and treatment for HIV+ people: A survey of Australian infertility clinics', *Australian and New Zealand Journal of Public Health*, vol 28, pp360–362

Platt, L. (2002) *Parallel Lives? Poverty Among Ethnic Minority Groups in Britain*, Child Poverty Action Group, London

Purcell, K., Schembri, M., Frazier, L. M., Rall, M. J., Shen, S., Croughan, M., Grainger, D. A. and Fujimoto, V. Y. (2007) 'Asian ethnicity is associated with reduced pregnancy outcomes after assisted reproductive technology', *Fertility and Sterility*, vol 87, pp297–302

Raleigh, V., Scobiue, A., Cook, A., Jones, S., Irons, R. and Halt, K. (2004) *Unpacking the Patient's Perspective: Variations in NHS Patient Experience in England*, Commission for Health Improvement, London

Richards, J. (2003) 'Ethical issues – the major faiths: A personal view', in D. Singer and M. Hunter (eds) *Assisted Human Reproduction*, Whurr Publishers, London

Roberts, R (1999) *Killing the Black Body: Race, Reproduction, and the Meaning of Liberty*, New York, Pantheon Books

Rodin, D. A., Bano, G., Bland, J. M., Taylor, K. and Nussey, S. S. (1998) 'Polycystic ovaries and associated metabolic abnormalities in Indian subcontinent Asian women', *Clinical Endocrinology*, vol 49, no 1, pp91–99

Roudsari, R. L., Allan, H. T. and Smith, P. A. (2007) 'Looking at infertility though the lens of religion and spirituality: A review of the literature', *Human Fertility*, vol 10, pp141–149

Rooij, F. B. van, van Balen, F and Hermanns, J. M. A.(2004) 'A review of Islamic Middle Eastern migrants: Traditional and religious cultural beliefs about procreation in the context of infertility treatment', *Journal of Reproductive and Infant Psychology*, vol 22, no 4, pp321–331

Rooij, F. B. van, van Balen, F. and Hermanns, J. M. A. (2006) 'Migrants and the meaning of parenthood: Involuntary childless Turkish migrants in the Netherlands', *Human Reproduction*, vol 21, pp1832–1838

Rooij, F. B. van, van Balen, F. and Hermanns, J. M. A. (2007) 'Emotional distress and infertility: Turkish migrant couples compared to Dutch couples and couples in Western Turkey', *Journal of Psychosomatic Obstetrics & Gynecology*, vol 28, no 2, pp87–95

Saffron, L. (2002) 'Can fertility service providers justify discrimination against lesbians?', *Human Fertility*, vol 5, pp42–46

Schenker, J. G. (2000) 'Women's reproductive health: Monotheistic religious perspectives', *International Journal of Gynecology & Obstetrics*, vol 70, pp77–86

Sharara, F. I. and McClamrock, H. D. (2000) 'Differences in in vitro fertilization (IVF) outcome between white and black women in an inner-city, university-based IVF program', *Fertility and Sterility*, vol 73, pp1170–1173

Smaje, C. and Field, D. (1997) 'Absent minorities? Ethnicity and the use of palliative care services', in D. Field, J. Hockey and N. Small (eds) *Death, Gender and Ethnicity*, Routledge, London

Smaje, C. and Le Grand, J. (1997) 'Ethnicity, equity and the use of health services in the

British NHS', *Social Science and Medicine*, vol 45, pp485–496

Smith, G. D. (2000) 'Learning to live with complexity: Ethnicity, socioeconomic position, and health in Britain and the United States', *American Journal of Public Health*, vol 90, no 11, pp1694–1698

Steinberg, D. L. (1997) 'A most selective practice. The eugenic logics of IVF' *Women's Studies International Forum*, vol 20, no 1, pp33–48

Stephen, E. H. and Chandra, A. (2000) 'Use of infertility services in the United States: 1995', *Family Planning Perspectives*, vol 32, no 3, pp132–137

White, L., McQuillan, J. and Greil, A. L. (2006) 'Explaining disparities in treatment seeking: The case of infertility', *Fertility and Sterility*, vol 85, pp853–857

Williams, D. R., Neighbors, H. W. and Jackson, J. S. (2003) 'Racial discrimination and health: Findings from community studies', *American Journal of Public Health*, vol 93, pp200–208

Woodward, B. J. and Norton, W. J. (2006) 'Lesbian intra-partner oocyte donation: A possible shake-up in the Garden of Eden?', *Human Fertility*, vol 9, pp217–222

Yebei, V. (2000) 'Unmet needs, beliefs and treatment-seeking for infertility among migrant Ghanaian women in the Netherlands', *Reproductive Health Matters*, vol 8, no 16, pp134–141

Yüksel, E. (1995) 'Kinderlosigkeit bei Turkischen Paaren: "Wie ein Baum ohne Fruchte" [Childlessness among Turkish couples; "Like a tree without fruit"], *TW Gy Gynakolongie*, vol 8, no 6, pp461–467

Zargar, A. H., Wani, A. I., Masoodi, S. R., Laway, B. A. and Salahuddin, M. (1997) 'Epidemiologic and etiologic aspects of primary infertility in the Kashmir region of India', *Fertility & Sterility*, vol 68, no 4, pp637–643

2

Infertility and Culture: Explanations, Implications and Dilemmas

Frank van Balen

Introduction

The concepts of infertility and childlessness are culturally moderated, and meanings and consequences differ in different social contexts. In the West, infertility, in the strict sense, is considered to be a medical term with a medical meaning (van den Akker, 2002). However, a number of other explanations and understandings of infertility exist cross-culturally. Not being able to have a child when one wishes has strong repercussions in many cultural contexts. However, at the same time, there is an increasing number of people who decide not to have children. Today, within Western societies, voluntary childlessness is increasingly respected and accepted as a 'lifestyle choice' (van Balen and Inhorn, 2002). Childlessness can therefore assume a number of guises: both voluntary and involuntary, medical and social. In this chapter, however, I will focus on involuntary childlessness. Involuntary childlessness can be caused by infertility, but is not strictly a medical concern. A person or couple can be defined as involuntarily childless for a number of social reasons including, in some contexts, the absence of children of a particular sex (Pashigian, 2002).

This chapter is concerned with the ways in which involuntary childlessness comes to be perceived in a particular cultural context, and correspondingly, how this impacts upon approaches to the treatment of infertility. The chapter begins with a brief overview of differing models of understanding infertility and its treatment, including a short history of the Western biomedical model. The chapter then discusses the consequences of childlessness among different societies and cultures, and compares the choices and possibilities for traditional and modern approaches to treatment, with a focus on the position and opportunities of those who have more recently migrated to the West.[1] Finally, the relationship between ethical dilemmas and culture is illustrated by a comparison of the stance of major religions towards in vitro fertilization (IVF), gamete donation and adoption.

Infertility: Explanatory models

As Helman (2000) has argued, cultural background has an important influence on many aspects of people's lives, including attitudes to illness. One important cultural difference regarding childlessness constitutes the 'explanatory model', the process by which illness is patterned, interpreted and treated (Kleinman, 1980). How does one explain the fact that a woman (or couple) cannot have children, and how does one conceive of a potential solution? A number of different models can be determined, each with different ways of understanding and dealing with involuntary childlessness: the biomedical model, the personalistic model and the naturalistic model, and these are described in relation to infertility below.

The biomedical model and the concept of 'infertility'

The dominant view of infertility in Western societies is that it is a medical condition and those infertile couples who wish to overcome their infertility most often (but not always) turn to the medical profession for a solution. The Western biomedical definition of infertility is of a reproductive impairment which is characterized by the lack of ability to conceive after one year of unprotected intercourse. Infertility is one of many social issues which became 'medicalized' in the course of the 20th century (Greil, 2002; Sandelowski and de Lacey, 2002). Historically, being unable to have children was medically described as 'sterility', an irreversible physical condition. Later, terms such as infertility and sub-fertility came into common usage, connoting a medically liminal state where reproductive capacity is in doubt. Sub-fertility, in particular, is a term that encompasses the reality that for many couples undergoing fertility treatment there remains a chance (albeit often very small) of conceiving a child spontaneously (van den Akker, 2002). This may be the case, for example, where infertility arises from malfunctioning ovaries, partial tubal blockage or inadequate semen. Diagnosis within the biomedical model can, however, be ambivalent, with gynaecological examinations sometimes resulting in contrasting findings. Where infertility is thought to arise from problems with the fallopian tubes for example, the findings of hysterosalpingography are not always confirmed by later laparoscopic investigation (Trimbos-Kemper, 1981).[2] Also, the results of semen analysis, both sperm count and sperm motility, are known to be variable, whether through real variation (Poland et al, 1986; Ombelet et al, 1998), or differences in measurement, expertise, criteria and methods of the laboratory and its personnel (Franken et al, 2000; Keel et al, 2000).

When psychosocial research into the causes of infertility began in the 1950s, psychoanalytic and psychodynamic theories were popular, and were used as the basis of understanding the aetiology of the phenomenon (Deutsch, 1947;

Benedek, 1952). At that time, knowledge about the mechanisms that caused 'sterility' was not very well developed. Diagnoses of low sperm count or sporadic ovulation could be made, but the underlying causes of these phenomena were not well established. Also, in around a third of cases, sterility was diagnosed as 'essential', 'functional' or 'psychosomatic', meaning that no physical cause could be diagnosed. Moreover, medically there was not much to be done about sterility. The most used 'treatment' was artificial insemination using donor semen where male sterility was suspected. This was a relatively effective treatment which was, however, shrouded in secrecy (Cohen, 2004).

In many cases, the cause of infertility was allegedly located in the mind–body interaction and considered psychosomatic. For instance, it was suggested that tubal blockage originated in tubal spasms, which in turn were evoked by the psyche. Deutsch (1947) and Benedek (1952) described frigidity, and anxiety about motherhood, as the cause of tubal spasms. Although the somatic causes of infertility have become increasingly clear in the past 50 years, psychogenic explanations have not disappeared completely and have often unintentionally served to increase rather than decrease the suffering of infertile women in particular (van Balen, 2002).

In the latter half of the 20th century, knowledge about infertility and its treatment progressed rapidly, beginning with the first clinical trials of the drugs to treat ovulatory problems in the 1960s. However, even by the turn of the century, about 10 per cent of infertility was considered to have an unknown cause (after extensive standardized examinations) (Verdurmen, 1997). The first successful attempt at IVF in 1978 marked the beginning of the 'new infertility' as a biological impediment to reproduction which could be bypassed, even if not 'cured', by biomedicine (Sandelowski and de Lacey, 2002). Alongside the rise in medical knowledge, more work was carried out in the socio-psychological realm. In these studies the term 'involuntary childlessness' arose, in order to accentuate the importance of the psychosocial context of infertility (van Balen and Inhorn, 2002). Through this definition, attention was focused on the consequences of childlessness, instead of on the causes. The word 'involuntary' was added to separate unwanted childlessness from the phenomenon of voluntary childlessness which received much attention during the second wave of feminism in the Western world (Gillespie, 1999; Thompson, 2002).

The biomedical response to infertility has been critiqued by a number of authors, particularly feminists, who have been concerned with the ways in which reproductive technologies such as IVF are invasive and present many ethical problems (Corea, 1987; Stanworth, 1987; Foster, 1995; Thompson, 2002). Social scientists have also expressed concerns about the fact that medical treatments offer only a partial solution to infertility, with large numbers of treatment cycles ending in failure (Becker, 2000; Sandelowski and de Lacey, 2002). The social, psychological and financial implications of IVF have been well documented (Greil, 1991; Becker, 2000; van Balen and Inhorn, 2002), with IVF being described by some as 'big business' (Pfeffer, 1993; Spar, 2006). It is important to acknowledge, therefore, that the biomedical response to infertility is an incomplete project and that up to half of couples without children do not seek treatment in the medical sphere (Greil, 1991).

The personalistic model

In the personalistic model of infertility, illness, or in this case the inability to conceive, is due to the purposeful active intervention of an agent. Somebody or something outside the woman (or couple) is assumed to be the cause of not having children. For example, these might be spirits, ghosts or humans with evil intentions. Often it is supposed that the woman has done something wrong which offended God(s), spirits or humans. Treatment is often sought by bringing offerings to Gods or other deities, visiting holy places and so on. Research by Nahar et al (1999) in Bangladesh and Liamputtong Rice (1999) in Laos demonstrates the relevance of personalistic models.

The naturalistic model

In the naturalistic model, illness is explained in impersonal, systemic terms rather than in terms of an active agent. However, the cause is not explained according to the biomedical infertility model, described above. Childlessness, for example, might be explained by not eating appropriate food, not being 'cold' or 'hot', having wrong 'fluids' or by some disequilibrium within the individual. Comparisons are made to natural problems such as barren fields and plants or trees without fruits (Yüksel, 1995; Kielman, 1998; Gerrits, 2002). Treatment is often given by herbalists or other 'traditional' therapies. However, in many poor-resource areas personalistic and naturalistic perspectives exist side by side or are combined in various folk models of infertility (Nahar, 2007; Inhorn, 2003).

Procreation beliefs

In addition to these perspectives on infertility, there are also culturally variable views of conception. In several parts of the world, beliefs exist that are fundamentally different from the Western biomedical model (Sundby, 1999; Inhorn and van Balen, 2002). For example, male monogenetic ideas of procreation whereby the complete 'pre-child' already exists in the sperm, and woman's contribution is limited to feeding and nurturing this pre-child in her womb, have existed in many cultures historically (Riddle, 1997; Stonehouse, 1999). In 1677, van Leeuwenhoeck thought he discovered the homunculus, a complete little being in the head of the sperm, when he used his lenses to investigate his own semen. Patriarchal monogenetic preformation models are also known as 'seed and soil' theories, whereby the male places the seed into the 'soil' of the female, who makes no contribution to the generative aspects of the foetus (van Rooij et al, 2004). Such beliefs, or a combination of these with aspects of the biomedical model are still in existence in several parts of the Middle East (Delaney, 1991; Inhorn, 1996; Inhorn, 2003).

Beliefs about procreation and the causes of infertility may also influence attitudes towards infertility treatments, though little is known about how this might impact on attitudes in the West (van Rooij et al, 2004). Furthermore, in poor-resource areas, Western treatments are often not available or are too expensive (van Balen and Inhorn, 2002), although the modern biomedical model is receiving more acceptance, especially among the urban higher and middle classes (Nahar, 2007).

In better resourced areas, many migrants come from areas where different explanatory models of procreation and infertility exist. Living in the Western world they will be confronted with the dominant biomedical model of the host society and its successes and failures. In contrast to the majority of people in poor-resource areas, biomedical infertility treatment is, in principle, within their reach and very often sought (Yüksel, 1995; van Rooij, 2008a), though in countries that lack a national healthcare reimbursement system, treatment is more difficult to obtain for the often economically disadvantaged migrants (Inhorn and Fakih, 2006).

In reality, the use of treatments based on personalistic and naturalistic models of infertility would often seem to be combined with biomedicine, suggesting that people hold complex sets of beliefs about the cause of their infertility. Yebei (1999), for example, found that infertile Ghanaian migrant women in the Netherlands, while accepting the assistance of biomedicine to solve their infertility, also attended healing meetings in the Netherlands and in Ghana, as some of them felt that their infertility was a consequence of witchcraft. At present, however, relatively little is known about the interplay between different explanatory models and procreation beliefs and the effect of these on treatment seeking, communication with physicians, choices for treatment, treatment compliance and effectiveness of treatment. It would also be interesting to know whether the knowledge and views that patients gather during their journey through the world of infertility investigations and treatments have an influence on the views and understanding of childlessness within the larger migrant communities and in the 'home' country.

Infertility treatment seeking among migrants

In a relatively short period of time, a number of different cultural groups have become established within Western states. There are, today, ethnic groups in most European countries that come from many parts of the world, including countries as diverse as Laos, Ghana, Chechnya, Iraq and Columbia. Migration has also brought great religious diversity to Western societies, with followers of religions and teachings such as Islam, Sikhism, Hinduism, Voodoo and Taoism now present. In all these religions, cultures and teachings, there exist different ideas about infertility, childlessness, possibilities for treatment, and its consequences (Schenker, 2000). These differences in culture and religion, while adding to the richness of life in a multicultural society, may also lead to strained and uneasy social and political relationships. There have been a number of public debates about the ethical acceptance of IVF and intracytoplasmic sperm injection (ICSI), the donation and use of gametes, pre-implantation diagnostics, and other issues around reproduction in the West (Thompson, 2002). However, these discussions have tended to be embedded within a Western perspective and have in general failed to incorporate views from non-Western viewpoints, as may be expressed by migrants from a range of cultural backgrounds.

Despite a lack of consultation and inclusion in public debate, for many migrants, modern biomedical treatment (such as IVF) is within reach. In many European countries at least, it is included in the national healthcare system, and costs are largely reimbursed, although the situation in the US is more complex (see Inhorn et al, Chapter 11). Also, there are indications that migrants increasingly know their way through the medical system and are able to find the appropriate treatment. Indeed, it is possible that they access treatment more effectively and at an earlier time than indigenous couples because of the greater pressure to have children (Eggert et al, 2007; van Rooij, 2008a). On the other hand, there may still be for some a gap between the biomedical world and the migrant culture as expressed, for example, in difficulties in communication and mutual understanding (see van Rooij and Korfker, Chapter 8). Couples who do not fluently speak the dominant language may be disadvantaged by inadequacies of interpreting services (Culley et al, 2006; Culley and Hudson, Chapter 6; van Rooij and Korfker, Chapter 8). Religious differences may also give rise to difficulties in the treatment process. However, little is known about the frequency of these occurrences, and the cultural linkages.

By comparison, in poor-resource areas from where some migrant groups are drawn, traditional healers are often the first choice of childless couples. This is because these treatments are in line with local explanatory models, and practised by people with the same culture and background of the childless couple. Moreover such treatment is often easily available and relatively cheap, compared with Western biomedicine (Sundby, 1997; Mogobe, 2000; Mariano, 2000; Gerrits, 2002; Nahar, 2007). Indeed for modern biomedicine it is often difficult to compete with traditional healing (Okonofua, 1996; Mulgaonkar, 2000; van Balen and Gerrits, 2001; Nahar 2007). High-tech treatment for infertility, like IVF and ICSI, is very expensive and only available in the big cities, if at all. Doctors are often of a different class or ethnic background from their patients, so they may speak in a different language or accent and belong to a different culture, making communication more difficult. Patients can often therefore feel a large distance between themselves and the physician. Finally, success rates of high-tech treatments are still not high, especially among newly qualified fertility specialists. There are many practical problems regarding equipment and electricity supply.

Culture and the consequences of childlessness

Though this book focuses on differences and diversity, it is also important to stress the commonalities in experiences of childlessness. Involuntarily childless women and men have existed and suffered in all times and all cultures. For any society and any culture, having children is an obvious prerequisite for its continuation. The importance of parenthood, and the demand to produce children, is almost universal. Not living up to this demand makes one an outsider, and people who cannot have children are in almost every culture considered as persons of less value than ones with children (van Balen and Inhorn, 2002). However, in no state or culture is the number of childless persons large enough to be a threat to its continuation. As a consequence, the problem of involuntary childlessness is often not socially recognized as important. Most politicians, for example, emphasize the role of families while neglecting those without children. In high-fertility, resource-poor areas especially, it is difficult for childless men and women to collectivize and to deploy power to advocate their interests and influence society (van Balen, 2002).

Though experiences vary, in the West consequences are mainly restricted to the individual and psychological realm. The biomedical model of understanding infertility means that couples in Western contexts often feel compelled to use highly invasive medical treatments to solve their infertility, which may have long-lasting emotional, financial and medical implications (Sandelowski, 1991). However, in poor-resource areas infertility often has even more severe social, legal, religious and economic consequences. Moreover these effects are gendered: women suffer more from these than men. Childlessness is also much more visible for a woman than a man. Women are generally the ones blamed for childlessness, and the ones who have the strongest negative experiences. Motherhood can be much more important for women than in Western culture (e.g. van Balen and Inhorn, 2002; Feldman-Savelsberg, 2002; van Rooij et al, 2004; Kagitcibasi and Ataca, 2005). Not being able to reach this status means having a very low social position, which often is accompanied by low self-esteem and feelings of being a 'failed woman'. Also, social isolation and stigma are frequently found (Inhorn, 1994; Okonofua et al, 1997; Kielman, 1998; Gerrits, 2002). There may be fear of abandonment among women (Dyer et al, 2002; Riessman, 2002), and in some cultures a man will resort to taking another wife (Inhorn, 1994; Sundby, 1999; Nahar, 2007). Childless women, and sometimes men, may be ridiculed and in extreme cases, abused by others (Sundby, 1997; Dyer et al, 2002; Dyer et al, 2004). In various cultures, childless people are banned from important social events, especially around marriage, childbirth, death or other '*rites de passage*' (Inhorn, 1994; Sundby, 1999; Gerrits, 2002). Also, in several non-Western religions, teachings, cults and ancestor worship are part of the rituals. Therefore, of course, it is important to have offspring. Children (especially sons) are supposed to care for their deceased parents in the 'afterlife'. Not having children makes people anxious about having a difficult time in the 'world beyond this one'. Infertile women may be accused of witchcraft. They may be seen to carry the 'evil eye', and best avoided, as described in some Arabian and West-African cultures (Inhorn, 1994; Meyer, 1994; Okonofua et al, 1997).

In her thesis about childlessness in Islamic Bangladesh, Nahar (2007) describes how in rural areas women without children are confined to their home, and are not allowed to do simple shopping and errands. In anthropological studies from the Indian subcontinent, we find narratives about the repayment of bride-gifts, and of childless women that feel used as slaves by their in-laws (family from the husband's side) to pay back 'their debt', that is, not producing a family heir (Bharadwaj, 1999; Riessman, 2002; Nahar, 2007). Moreover, childless women are supposed to have 'spare time' to work for their in-laws. At the same time, it is also noted that even within these cultural restraints, women are able to enact agency and 'resist' labelling and stigma to some extent, depending on their social and economic circumstances (Remennick, 2000; Riessman, 2000).

Based on comparisons with studies done in non-Western areas, one might expect greater negative effects of childlessness among migrants in the West, especially among women; however, this area is significantly under-researched. It is also important to note that attitudes to childlessness and to treatment options will be influenced by acculturation processes (see also van Rooij et al, 2006). While couples are undoubtedly likely to be influenced by the cultures of their countries of origin, it cannot be automatically assumed that they will be the same, especially in the case of long-established minority ethnic communities. Other contributions in this volume (Inhorn et al; Culley and Hudson) also demonstrate the importance of non-cultural factors in accessing fertility treatment in the West.

Religion, solutions to infertility and ethical dilemmas

If one cannot have children in the 'normal way', there are various ways to solve involuntary childlessness, which can often give rise to moral dilemmas for infertile couples. In many cases people seek answers to these questions within their culture, and look for guidance, especially in their religion. Indeed, authorities within the three major monotheistic religions have taken specific positions on alternative solutions to infertility (Schenker, 1992). Moreover, knowledge of these religions is often limited within the medical establishment. As an illustration, I will make a short comparison of the stance of three major monotheistic beliefs towards three frequently used alternative ways to have a child: IVF, gamete donation and adoption.

Modern forms of assisted conception became possible following the application of new knowledge of reproductive endocrinology in the 1960s. The advent of IVF in 1978 gave raise to public debate regarding the ethical and moral issues involved in bringing about human fertilization outside the body. Religious commentators in particular expressed widespread concern about the 'unnatural' nature of IVF and the fact that the medical profession were 'playing God' (Schenker, 2005). The Roman Catholic Church is still very much opposed to IVF because it breaks the 'inseparable link' between sex and procreation. Cardinal Ratzinger, past president of the Congregation for the Faith, the institution that is supposed to preserve faith and give official answers to the controversies of the day, spoke strongly against IVF, as   expressed in the Instruction Donum Vitae (Congregation for the Faith, 1987). At a conference on reproductive techniques he explicitly and unconditionally denounced and forbade IVF (Rensen, 1987). Today Ratzinger is Pope Benedictus XVI, and the viewpoint expressed in 1987 is still valid. In contrast with Roman Catholicism, various denominations of Protestantism have liberal attitudes towards IVF with the couples' own gametes (Dutney, 2007). Mainstream Islam (both Sunni and Shia), accept IVF and related techniques of assisted conception, on the grounds that God has given humans the power to use their intelligence and skills to help other people (Husain, 2000). Various Islamic scholars have issued fatwas permitting the use of IVF as long as the sperm and the egg are from husband and wife. The fatwas of the Al Ahzar University of Cairo have played a great role in accepting this technology (Serour, 1996; Serour, 2002). Also, orthodox and ultra orthodox Judaism accepts IVF completely. Based on this position, the state of Israel reimburses IVF, for an infinite number of treatments (Kahn, 2002).

Regarding the acceptability of donated gametes, whether sperm or eggs, Sunni Islam and the Roman Catholic Church are both vehemently opposed (Congregation for the Faith, 1987; Serour, 1996). However, various authoritative Shiite Muslim clerics accept the opportunity of having a short-term 'extra marriage' with an ovum-donating woman, thus making IVF with egg donation acceptable (Tremayne, 2006). However, the egg-receiving wife is not considered by most clerics as the mother in the legal sense. Because women are not allowed to marry more than one husband, a comparable solution regarding sperm donation is not possible (Jafarzadeh, 2000) and this remains prohibited (Serour, 2002; Inhorn, 2006). Regarding donated gametes, ultra orthodox Jews take what might be for non-believers a surprising position. Certain rabbinic interpretations do not allow donation of sperm by another Jew, as it is considered a form of adultery. In contrast, the majority of rabbis allow sperm donation from a Goy (non-Jewish man). In their Halakhic definition, adultery is confined to sexual intercourse of a married Jewish woman with a Jewish man that is not her husband. However, there is one exception: in the case of the 'Kohen' (priest class), this solution to infertility is not accepted, because of the importance of the male lineage for this group (Kahn, 2002). Also, because Jewishness is conferred through the female line, the issue of egg donation is more complex. Many rabbinical authorities reject this, but some permit it as long as the husband consents (Kahn, 2002).

With respect to adoption, Islam does not allow adoption, though Islamic texts are very positive about helping orphans. Western style adoption or adoption according to Roman law, in which the adopted child becomes a full member of his new family with all rights attached, such as name and heritage, is forbidden. This is because in the Koran it is described that Mohammed himself repudiated the former adoption of his already grown up and married son (Zeid). This change was ordered through a revelation to Mohammed (Koran, 33:1–9) in which he was told that blood relations are closer than any other. Thereafter, Mohammed married the wife of Zeid (Zeinab), by which Allah made it clear that this is a legitimate action for true believers (Koran, 33:36–40). In contrast to Islam, the Roman Catholic Church accepts adoption. Also, in the Jewish tradition, adoption is condoned; however, ultra orthodox Jews stress the importance of a Jewish lineage of the child, and a conversion ceremony is needed (Jewish Adoption, 2008).

Of course, there are many more religious and cultural differences regarding the many options and varieties of treatment and solving childlessness. In dealing with treatment and other options to solve childlessness, one should be aware of the religious beliefs and cultural attitudes of infertile individuals, although it is important to recognize that people may be more or less religiously devout or observant. There have been relatively few studies that empirically investigate the impact of religious beliefs and teachings on infertility treatment either in terms of whether religious teachings are followed by patients or whether couples experience stress if they undertake religiously prohibited treatments. It is often assumed that the teachings of Islam have much more influence on the daily life of believers than mainstream Protestantism and Catholicism. However, studies with Turkish migrants in Germany revealed that about 20 per cent of them would accept donor insemination, when this is the only solution possible (Yüksel, 1995). Also, van Rooij discovered that some Muslim migrants in the Netherlands secretly used donation to have a child (van Rooij, 2008b). Van Balen found that Roman Catholics in the Netherlands did not follow the instructions of the church (van Balen, 1991), as the percentage of Roman Catholics among those using IVF was about the same as the national average. However, in this respect one has to recognize that the power of the Roman Catholic Church on people's behaviour – at least in North West Europe – has dwindled considerably (see Cotter, Chapter 10). There are, however, some studies from non-Western societies in particular that demonstrate that for some groups religion plays a role in 'healing' infertility, and women in particular often get help for infertility concurrently from religious/spiritual and biomedical sources (Roudsari et al, 2007) and this may well be the case for migrant communities.

Conclusion

As society becomes more diverse it is important to become aware of the different meanings and different consequences of not having children among different cultural groups. Though there are large differences among, and indeed, within the various migrant groups, childlessness is for most migrants an extremely feared condition. Also, the importance of motherhood still appears to be paramount. Involuntarily childless migrants may have to come to terms with conflicting explanatory models about the origin, causes and solutions of infertility. In the dilemmas around treatment seeking it is important to recognize the significance of religious guidance and cultural constraints. It is important to stress, however, that ultimately, the decisions are made by the couples themselves.

Notes

1   I will reserve the term 'childlessness' for discussing the psychological and social impact of not being able to have children; allowing a broader definition than is permitted by use of the term infertility.
2   Hysterosalpingography is a contrast X-ray picture of dye being injected into the Fallopian tubes, used to diagnose tubal blockage. Laparoscopy is an investigation by the eye or monitor through a tube inserted in the inflated abdomen (see Glossary).

References

Akker, O. van den (2002) *The Complete Guide to Infertility. Diagnosis, Treatment, Options*, Free Association Books, London

Balen, F. van (1991) *Een Leven zonder Kinderen: Beleving, Stress en Aanpassing [A Life without Children: Experience, Stress and Adaptation]*, Dekker and van de Vegt, Assen, the Netherlands

Balen, F. van and Gerrits, T. (2001) 'Quality of infertility care in poor-resource areas and the introduction of new reproductive technologies', *Human Reproduction*, vol 16, pp215–219

Balen, F. van (2002) 'The psychologization of infertility', in M. C. Inhorn and F. van Balen (eds) *Infertility around the Globe: New Thinking on Childlessness, Gender and Reproductive Technologies*, University of California Press, Berkeley, CA

Balen, F. van and Inhorn, M. C. (2002) 'Interpreting infertility: A view from the social sciences', in M. C. Inhorn and F. van Balen (eds) *Infertility around the Globe, New Thinking on Childlessness, Gender and Reproductive Technologies*, University of California Press, Berkeley, CA

Becker, G. (2000) *The Elusive Embryo: How Women and Men Approach New Reproductive Technologies*, University of California Press, Berkeley, CA

Benedek, T. (1952) 'Infertility as a psychosomatic defence', *Fertility and Sterility*, vol 3, pp80–86

Bharadwaj, A. (1999) 'Infertility and gender: A perspective from India', in F. van Balen, T. Gerrits and M. Inhorn (eds) *Social Science Research on Childlessness in a Global Perspective*, University of Amsterdam, Amsterdam, pp65–74

Cohen, S. (2004) 'The invisible man. Artificial insemination by donor and the legislation on donor anonymity: A review', *Journal of Family Planning and Reproductive Health Care*, vol 30, no 4, pp270–273

Congregation for the Faith (1987) 'Instruction Donum Vitae, on respect for the beginning of life and the value of procreation', Polyglot Press, Vatican

Corea, G. (1987) 'What the king cannot see', *Women's Health*, vol 13, nos 1–2, pp77–93

Culley, L. A., Hudson, N., Rapport, F. L., Katbamna, S. and Johnson, M. R. D. (2006) 'British South Asian communities and infertility services', *Human Fertility*, vol 9, no 1, pp37–45

Delaney, C. (1991) *The Seed and the Soil. Gender and Cosmology in a Turkish Village*, UCLA-press, Berkeley, CA

Deutsch, H. (1947) *The Psychology of Women: A Psychoanalytical Interpretation. Vol. 2. Motherhood*, Grune and Straton, London, and Research Books, New York

Dutney, A. (2007) 'Religion, infertility and assisted reproductive technology', *Best Practice & Research Clinical Obstetrics & Gynaecology*, vol 21, no 1, pp169–180

Dyer, S. J., Abrahams, N., Hoffman, M. and van der Spuy, Z. M. (2002) '"Men leave me as I cannot have children": Women's experiences with involuntary childlessness', *Human Reproduction*, vol 17, pp1663–1668

Dyer, S. J., Abrahams, N., Mokoena, N. E. and van der Spuy, Z. M. (2004) '"You are a man because you have children": Experience, reproductive health knowledge and treatment-seeking behaviour among men suffering from couple infertility in South Africa', *Human Reproduction*, vol 19, pp960–967

Eggert, J., Li, X. and Sundquist, K. (2007) 'Country of birth and hospitalization for pelvic inflammatory disease, ectopic pregnancy, endometriosis, and infertility: A nationwide study of 2 million women in Sweden', *Fertility and Sterility*, available online 19 September 2007

Feldman-Savelsberg, P. (2002) 'Is infertility an unrecognized public health and population problem? The view from the Cameroon Grassfields', in M. C. Inhorn and F. van Balen (eds) *Infertility around the Globe: New Thinking on Childlessness, Gender and Reproductive Technologies*, University of California Press, Berkeley, CA

Foster, P. (1995) *Women and the Health Care Industry: An Unhealthy Relationship?*, Open

University Press, Milton Keynes

Franken, D. R., Smith, M., Menkveld, R., Kruger, T. F., Sekkade-Kigondu, C., Mbizvo, M. and Akande, E. O. (2000) 'The development of a continuous quality control programme for strict sperm morphology among sub-Saharan African laboratories', *Human Reproduction*, vol 15, pp667–671

Gerrits, T. (2002) 'Infertility and matrilineality: The exceptional case of the Macua of Mozambique', in M. C. Inhorn and F. van Balen (eds) *Infertility around the Globe: New Thinking on Childlessness, Gender and Reproductive Technologies*, University of California Press, Berkeley, CA

Gillespie, R. (1999) 'Voluntary childlessness in the United Kingdom', *Reproductive Health Matters*, vol 7, no 13, pp43–53

Greil, A. L. (1991) *Not Yet Pregnant: Infertile Couples in Contemporary America*, Rutgers University Press, New Brunswick, NJ

Greil, A. L. (2002) 'Infertile bodies: Medicalization, metaphor and agency', in M. Inhorn and F. van Balen (eds), *Infertility around the Globe: New Thinking on Childlessness, Gender and Reproductive Technologies*, University of California Press, Berkeley, CA

Helman, C. G. (2000) *Culture, Health and Illness*, Edward Arnold, London

Husain, F. A. (2000) 'Reproductive issues from the Islamic perspective', *Human Fertility*, vol 3, pp124–128

Inhorn, M. C. (1994) *Quest for Conception, Gender, Infertility and Egyptian Medical Traditions*, University of Pennsylvania Press, Philadelphia, PA

Inhorn, M. C. (1996) *Infertility and Patriarchy: The Cultural Politics of Gender and Family Life in Egypt*, University of Pennsylvania Press, Philadelphia, PA

Inhorn, M. C. (2003) *Local Babies, Global Science: Gender, Religion and In Vitro Fertilization in Egypt*, Routledge, London

Inhorn, M. C. (2006) 'Making Muslim babies: IVF and gamete donation in Sunni vs Shia Islam', *Culture, Medicine and Psychiatry*, vol 30, pp427–450

Inhorn, M. C. and Fakih, M. H. (2006) 'Arab Americans, African Americans, and infertility: Barriers to reproduction and medical care', *Fertility and Sterility*, vol 85, pp844–852

Inhorn, M. C. and van Balen, F. (eds) (2002) *Infertility around the Globe: New Thinking on Childlessness, Gender and Reproductive Technologies*, University of California Press, Berkeley, CA

Jafarzadeh, M. (2000) 'Iranian-law attitude to assisted reproduction: Present situation and future perspectives', paper presented during the ESHRE conference, Italy

Jewish Adoption (2008) www.adoptioninformation.com/Jewish_adoption, accessed 20 March 2008

Kagitcibasi, C. and Ataca, B. (2005) 'Value of children and family change: A three-decade portrait from Turkey', *Applied Psychology International Review*, vol 54, pp317–337

Kahn, S. M. (2002) 'Rabbis and reproduction: The use of new reproductive technologies among ultraorthodox Jews in Israel', in M. C. Inhorn and F. van Balen (eds) *Infertility around the Globe: New Thinking on Childlessness, Gender and Reproductive Technologies*, University of California Press, Berkeley, CA

Keel, B. A, Quinn, P., Schmidt, C. F., Serafy Jr, N. T. and Schalue, T. K. (2000) 'Results of the American Association of Bioanalysts national proficiency testing program in andrology', *Human Reproduction*, vol 15, pp680–686

Kielman, K. (1998) 'Barren ground: Contesting identities of infertile women in Pemba, Tanzania', in M. Lock and P. A. Kaufert (eds) *Pragmatic Women and Body Politics*, Cambridge University Press, Cambridge

Kleinman, A. (1980) *Patients and Healers in the Context of Culture: An Exploration of the Borderland between Anthropology, Medicine, and Psychiatry*, University of California

Press, Berkeley, CA

Koran, The, Fourth revised edition (1974) Translated with notes by N. J. Dawood, Penguin Books, Harmondsworth, UK

Liamputtong Rice, P. (1999) 'A barren body: The cultural interpretation of infertility among Hmong women in Australia', in F. van Balen, T. Gerrits and M. Inhorn (eds) *Social Science Research on Childlessness in a Global Perspective*, University of Amsterdam, Amsterdam

Mariano, E. (2000) 'Conceptions about woman's childlessness and land's infertility in a Rhonga community in the South of Mozambique', in F. van Balen, T. Gerrits and M. Inhorn (eds) *Social Science Research on Childlessness in a Global Perspective*, University of Amsterdam, Amsterdam

Meyer, B. (1994) 'Satan, slangen en geld [Satan, snakes and money]', in H. Driessens and H. de Jonge (eds) *In de Ban van de Betekenis, Proeven van Symbolische Antropologie [In the Spell of Meaning, Essays in Symbologic Anthropology]*, Sun, Nijmegen

Mogobe, K. (2000) 'Traditional therapies for infertility: Implications for the brokerage role of the nurse in Botswana', in F. van Balen, T. Gerrits and M. Inhorn (eds) *Social Science Research on Childlessness in a Global Perspective*, University of Amsterdam, Amsterdam

Mulgaonkar, V. (2000) *A Research and An Intervention Program on Women's Reproductive Health in Slums of Mumbai*, Sujeevan Trust, Mumbai

Nahar, P. (2007) *Childless in Bangladesh: Suffering and Resilience among Rural and Urban Women*, University of Amsterdam, Amsterdam

Nahar, P., Sharma, A., Sabin, K., Baqui, A., Begum, L. and Assam, K. (1999) 'An explanatory model of infertility among the urban population of Bangladesh', in F. van Balen, T. Gerrits and M. Inhorn (eds) *Social Science Research on Childlessness in a Global Perspective*, University of Amsterdam, Amsterdam

Okonofua, F. E. (1996) 'The case against new reproductive technologies in developing countries', *British Journal of Obstetrics and Gynecology*, vol 103, pp957–962

Okonofua, F. E., Harrus, D., Odebiyi, A., Kane, T. and Snow, R. C. (1997), 'The social meaning of infertility in Southwest Nigeria', *Health Transition Review*, vol 7, pp205–220

Ombelet, W., Bosmans, E., Janssen, M., Cox, A., Maes, M., Punjabi, U., Blaton, V., Gunst, J., Haidl, G., Wouters, E., Spiessens, C., Bornman, M. S., Pienaar, E., Menkveld, R. and Lombard, C. (1998) 'Multicenter study on reproducibility of sperm morphology assessments', *Archives of Andrology*, vol 4, pp103–114

Pashigian, M. J. (2002) 'Conceiving the happy family: Infertility and marital politics in Northern Vietnam', in M. Inhorn and F. van Balen (eds), *Infertility around the Globe: New Thinking on Childlessness, Gender and Reproductive Technologies*, University of California Press, Berkeley, CA

Pfeffer, N. (1993) *The Stork and the Syringe: A Political History of Reproductive Medicine*, Polity Press, Cambridge

Poland, M. L., Giblin, P. T., Ager, J. W. and Moghisse, K. D. (1986) 'Effects of stress on semen quality in semen donors', *International Journal of Fertility*, vol 31, pp229–231

Remennick, L. (2000) 'Childless in the land of imperative motherhood: Stigma and coping among infertile Israeli women', *Sex Roles*, vol 43, no 11, pp821–841

Rensen, R. (1987) 'Ratzinger hekelt "kinderproduktie"' [Ratzinger lashes 'child production'], *NRC-Handelsblad*, 11 March, p5

Riddle, J. M. (1997) *Eve's Herbs: A History of Contraception and Abortion in the West*, Harvard University Press, Cambridge, MA

Riessman, C. K. (2000) 'Stigma and everyday practices: Childless women in south India', *Gender & Society*, vol 14, no 1, pp111–135

Riessman, C. K. (2002) 'Positioning gender identity in narratives of infertility: South Indian women's lives in perspective', in M. C. Inhorn and F. van Balen (eds) *Infertility around the Globe: New Thinking on Childlessness, Gender and Reproductive Technologies*, University of California Press, Berkeley, CA

Rooij, F. B. van (2008a) 'The experience of involuntarily childless Turkish migrants: Parenthood motives, psycho-social consequences, responses and help-seeking behavior', PhD thesis, SCO- Kohnstamm Instituut, University of Amsterdam, Amsterdam

Rooij, F. B. van (2008b) personal communication

Rooij, F. B. van, van Balen, F. and Hermanns, J. M. A. (2004) 'A review of Islamic Middle Eastern migrants: Traditional and religious cultural beliefs about procreation in the context of infertility treatment', *Journal of Reproductive and Infant Psychology*, vol 22, pp321–331

Rooij, F. B. van, van Balen, F. and Hermanns, J. M. A. (2006) 'Migrants and the meaning of parenthood: Involuntary childless Turkish migrants in the Netherlands', *Human Reproduction*, vol 21, no 7, pp1832–1838

Roudsari, R. L, Allan, H. T. and Smith, P. A. (2007) 'Looking at infertility through the lens of religion and spirituality: A review of the literature', *Human Fertility*, vol 10, no 3, pp141–149

Sandelowski, M. (1991) 'Compelled to try: The never-enough quality of conceptive technology', *Medical Anthropology Quarterly*, vol 5, no 1, pp29–47

Sandelowski, M. and de Lacey, S. (2002) 'The uses of "disease": Infertility as a rhetorical vehicle', in M. Inhorn and F. van Balen (eds) *Infertility around the Globe, New Thinking on Childlessness, Gender and Reproductive Technologies*, University of California Press, Berkeley, CA

Schenker, J. G. (1992) 'Religious views regarding treatment of infertility by assisted reproductive technologies', *Journal of Assisted Reproduction and Genetics*, vol 9, no 1, pp3–8

Schenker, J. G. (2000) 'Women's reproductive health: Monotheistic religious perspectives', *International Journal of Gynecology & Obstetrics*, vol 70, pp77–86

Schenker, J. G. (2005) 'Assisted reproductive practice: Religious perspectives', *Reproductive Biomedicine Online*, vol 10, pp310–319

Serour, G. I. (1996) 'Bioethics in reproductive health: A Muslim perspective', *Middle East Fertility Society Journal*, vol 1, pp30–35

Serour, G. I. (2002) 'Attitudes and cultural perspectives on infertility and its alleviation in the Middle East Area', in E. Vayena, P. J. Rowe and P. D. Griffen (eds) *Current Practices and Controversies in Assisted Reproduction*, World Health Organization, Geneva

Spar, D. L. (2006) 'Where babies come from: Supply and demand in an infant marketplace', *Harvard Business Review*, vol 4, no 2, pp133–142

Stanworth, M. (ed) (1987) *Reproductive Technologies: Gender, Motherhood and Medicine*, Polity Press/Basil Blackwell, Cambridge

Stonehouse, J. (1999) 'Procreation, patriarchy and medical science: Resistance to recognizing maternal contributions in European embryological thought', in P. Loizos and P. Heady (eds) *Conceiving Persons: Ethnographies of Procreation, Fertility, and Growth*, Athlone Press, London, New Brunswick, NJ

Sundby, J. (1997) 'Infertility in the Gambia: Traditional and modern health care', *Patient Education and Counseling*, vol 31, pp29–37

Sundby, J. (1999) 'Family formation, norms and infertility care perspectives from the North and South', in F. van Balen, T. Gerrits and M. Inhorn (eds) *Social Science Research on Childlessness in a Global Perspective*, University of Amsterdam, Amsterdam

Thompson, C. M. (2002) 'Fertile ground: Feminists theorize infertility', in M. C. Inhorn and F. van Balen (eds) *Infertility around the Globe, New Thinking on Childlessness,*

Gender and Reproductive Technologies, University of California Press, Berkeley, CA

Tremayne, S. (2006) 'Not all Muslims are Luddites', *Anthropology Today*, vol 22, no 3, pp1–2

Trimbos-Kemper, G. C. M. (1981) *Tubachirugie [Tubal surgery]*, University of Leiden, Leiden

Verdurmen, J. (1997) 'Keuzes bij onvruchtbaarheid: besluitvormingsprocessen bij onvruchtbare paren [Choices around infertility: decision-making among infertile couples]', PhD thesis, University of Amsterdam, Amsterdam

Yebei, V. N. (1999) 'Infertility management strategies among infertile migrant Ghanian women in Amsterdam', in F. van Balen, T. Gerrits and M. Inhorn (eds) *Social Science Research on Childlessness in a Global Perspective*, University of Amsterdam, Amsterdam

Yüksel, E. (1995) 'Kinderlosigkeit bei Türkischen Paaren: "Wie ein Baum ohne Fruchte" [Childlessness among Turkish couples: "Like a tree without fruit"]', *TW Gynakologie*, vol 8, pp461–467

3

Making Sense of Ethnic Diversity, Difference and Disadvantage within the Context of Multicultural Societies

Karl Atkin

Advanced industrial societies struggle to provide accessible and appropriate healthcare for culturally and linguistically diverse populations (Healy and McKee, 2004). These are long-standing and familiar problems, but our failure to engage with them contributes to flawed understandings, inappropriate responses and wasted resources (Atkin and Chattoo, 2007). Taking this as a starting point, the purpose of this chapter is to provide a general analytical framework, in which to make sense of the emerging debate on ethnicity, fertility and reproductive technologies. I make no apologies for its broad scope. There is little discussion and much less research exploring ethnicity and fertility (National Institute for Clinical Excellence, 2004), making it difficult to establish a credible evidence base on which to base interventions (Balen and Crawshaw, 2006) and, perhaps more importantly, creating the potential, as debates about ethnicity and fertility develop, to 'reinvent the wheel' and not learn from previous empirical and theoretical insights (Bhopal, 2007).

The chapter specifically offers an agenda for future engagement, in which social justice assumes prominence. My concern is not to offer neat prescriptive cultural descriptions, purporting to explain and manage 'ethnicity', but to offer a more general discussion that contextualizes diversity and difference, without recourse to simplistic explanations and naive solutions, which perpetuate disadvantage and discrimination. In practical terms, I begin by exploring what is meant by diversity and difference and the implications of such debates for how we come to 'imagine' multicultural societies. I then offer a more specific account of institutional racism, which has assumed recent legitimacy in explaining disadvantage and discrimination in the UK, and examine its relevance in understanding the experience of ethnic minority populations in multicultural societies more generally. I end with a discussion of how we can best use evidence to improve outcomes, while acknowledging that policy and practice are sometimes better at documenting the implications of ethnic diversity, rather than ensuring they do not become the basis of inequalities.

This is why this account is prefaced by reflecting on the dilemmas facing those attempting to accommodate ethnic diversity, within the context of a multicultural society. These emerge as fundamental in making sense of this narrative, providing a coherent expression of intent. I accept there is a depressing familiarity to many of the issues raised. The well understood processes of disadvantage and discrimination can sometimes delude those who are familiar with the complexity of current debates into thinking that there is little else to do. Our growing awareness, however, has not always equated with a more responsive welfare provision. The UK offers several good examples of this. Empirical research has long discredited stereotypes of minority ethnic families who 'look after their own' (see Cameron et al, 1989). These ideas, however, continually surface in the attitudes of many practitioners (Atkin and Chattoo, 2007). Consequently, familiarity – taken for granted by some – is not always evident in the research or practice of others. Essentialism, which has long been discredited in academic circles, is beginning to re-emerge, as research, policy and practice slowly respond to the multicultural nature of society (Chattoo and Ahmad, 2008). Debates about the complex nature of ethnic identity seem cast aside, as mainstream research, unfamiliar with such debates, applies idiosyncratic and poorly contextualized ideas of the meaning of ethnicity (Craig, 2007). The supposed fatalism of South Asian patients with chronic illness, for example, is a common feature of the nursing and medical literature, with little recognition that fatalism does not preclude active engagement with the condition (Atkin and Ahmad, 2000) or that fatalism can occur, irrespective of ethnicity, as people try to make sense of what has happened to them (Chattoo and Ahmad, 2004).

Identity, diversity and difference

Making sense of ethnic diversity and difference requires going beyond conceptualizing ethnicity as a straightforward marker of disadvantage and discrimination – in which institutional cultures and racisms sustain inequalities – to emphasizing the contingent process through which ethnicity finds expression in social relationships. This should be the starting point of any discussion about ethnic diversity, as it enables us to understand ethnicity within its broader social and cultural context, without any preconceptions. Knowing who we are and seeking legitimacy for this understanding of ourselves is at the heart of social life, and a person's ethnic identity is a fundamental part of this process (Jenkins, 1996). Ethnic identity, however, is notoriously difficult to define and conceptual confusion is commonplace (Bradby, 2003), making cross-country comparisons especially difficult (see Johnson and Borde, Chapter 4). This represents the first problem. Nor is ethnic identity merely the preserve of ethnic minorities, as is sometimes assumed in the literature. Everyone has an ethnic origin, and its universality is equally important to the idea's meaning and legitimacy.

In some ways, conceptual confusion is understandable and reflects the essence of ethnic identity. The multifaceted nature in which we have come to understand 'ethnic' identity has many advantages (Atkin and Chattoo, 2007). If nothing else, it reminds us of the complex and shifting nature of ethnicity, as it comes to embody language, religion, faith, culture, ancestry, nationality, histories of migration and a shared heritage (Ahmad et al, 2002). Such embodiments – and the subsequent discursive practices they engender – become fundamental to how we 'imagine' ethnic identity. Further, these embodiments generate complex relationships, occurring in time and space, negotiated according to how the different aspects of ethnic identity support, sustain, reinforce and contradict each other. For many South Asian people living in the UK, for example, religion and ethnicity often reinforce each other, especially among first generation migrants, though this does not seem the case for African-Caribbean young people for whom the importance of religious identity is less strong compared to their parents (Modood et al, 1997). The detailed nuances of such discussions, however, perhaps mask a broader point. A person's sense of ethnicity offers cultural resources and a defined framework, which can be employed creatively to make sense of identity (Hall, 1996).

As part of this, ethnic identity is increasingly seen as a political symbol, defining not just exclusion by a powerful majority but also a source of pride and belonging (Parekh, 2006): in other words, a mobilizing resource, which enables minority ethnic populations, as an 'imagined' moral community, to celebrate their difference and make legitimate demands as citizens (Anderson, 2006). A recent example of this is how for some Muslims living in Western Europe, religion represents more a political statement than an expression of personal faith, responding to the growing politicization of Islam within Western countries (Craig, 2007).

Such discussion, however, reminds us about another important aspect of ethnicity: identities become negotiated within the context of social relationships. Who you are, is as much about how others come to define you, as your own perception of identity (see Bourdieu, 1977). The labels associated with 'ethnic minority' are, for example, often assigned by the majority ethnic population. Current census categories in the UK reflect Britain's colonial past, confusing skin colour, race and ethnicity and geography. Similar observations could be made about how other European, North American and Australasian states define ethnicity. State practices further reinforce this dynamic, creating the more pervasive space in which diversity and difference become negotiated, above and beyond the individual's own identifications. Three distinct – and possibly contradictory – premises become evident, representing the 'Janus-faced' nature of policy discourses (see Craig, 2007).

This is manifest in the UK, which has specific legislation that aims to tackle institutional racism (see below). The legislation, introduced in April 2001 and representing an amendment to the 1976 Race Relations Act, made all statutory agencies responsible for promoting equal opportunities and identifying and tackling institutional racism within their organizations. This reflected a willingness on the part of the British state to tackle disadvantage, although its enactment is far from straightforward. Second – and in response to anxieties about the willingness and ability to integrate of ethnic minority populations (see Ouseley, 2001) and a suggested existential crisis about how people in the UK define themselves (Royal Society of Arts, 2006) – the UK government has pursued a policy of community cohesion, aiming to facilitate a national identity and a more inclusive society, where a shared sense of 'Britishness' is offered as a means of managing a multicultural society (Home Office, 2004). Community cohesion certainly recognizes the importance of empowering communities and tacking inequalities, although there remains an unresolved tension. To what extent can a defined sense of 'Britishness' appropriately accommodate diversity and difference? Or will it be used to remind ethnic minorities of their 'otherness', in a way that pathologizes their experience? (see Atkin and Chattoo, 2007). A ritualized enactment of Britishness could, for example, easily become associated with a moral and cultural superiority, which has little respect for cultural difference (see Judd, 1996). Third, it should not be forgotten that discussion about diversity and difference occurs against a repressive and restrictive approach to immigration, which is longstanding and implies that ethnic diversity is a problem, harmful to the British state (Williams, 1996).

Other post-industrial states have similar contradictions within their policies and practices, where measures to tackle discrimination occur alongside strategies to create a sense of national identity, as well as legislation which clearly defines who can enter the country and be regarded as a citizen of that country (see Bleich, 2003, for a specific example and Kymlicka, 2001, for a more general discussion). In some countries, such as North America, Australia and New Zealand, the situation is further complicated because some populations classified as 'ethnic minority' also happen to be the original indigenous population, and this raises fundamental questions about the relationship between citizenship and national identity. As we have seen, however, such debates represent an inherent and ongoing tension in how 'ethnicity' is imagined and defined in different multicultural societies: ethnicity is produced within a particular social and historical context. I now explore this in more detail.

Responding to ethnic identity

Ethnic minority populations living in the UK remain incredibly diverse in the cultural values, religions followed, languages spoken, history of migration and country of origin, with similarities between and within different groups (Gunaratnam, 1997). This is similar to other multicultural societies. Where does this, however, leave us analytically? This account suggests that ethnic identities are contingent because of their responsiveness to context (Giddens, 1991). Consequently, ethnic identities, in their broadest sense, become cultivated and negotiated within a variety of structures and ideological frameworks (Hall, 1996). Such identities have different meanings to different people, in different contexts. In some instances a person might wish to emphasize religious identity; in others, the sense of national heritage might be more important. Nor are these identities immutable but will themselves change over time and with changing circumstances (see Ahmad et al, 2000).

The idea that ethnic identities are flexible is, of course, not new (Papastergiadis, 1982). Age, generational relations, history of migration, social class, gender, sexuality, ethnicity and religion represent important identifications, the inter-relationships of which are complex and connected to questions of power, structure and history (Ahmad et al, 2002). This is why we should guard against allowing ill-informed, generalized cultural stereotypes to shape the development of policy and practice (Nizalova, 2000). Ethnic identity carries an implicit danger of becoming stripped of context, subject to stereotypes and mythologies, rather than an expression of a person's negotiation of multiple identities within specific social and political contexts (see Bauman, 1992). Ethnic identity is not something fixed or predefined. It is – as we have seen – produced and negotiated within a particular social context, which embodies individual interpretation and preferences in relation to the values of significant others. Ethnic identity emerges as flexible, negotiable and contingent: a resource, which could be used creatively to support a person's sense of who he/she is, while simultaneously sustaining disadvantage and discrimination.

In more practical terms, current debates struggle to engage with the relationship between ethnicity, socio-economic status, age and gender, the outcome of which can obscure fundamental similarities as well as differences among populations (Smaje, 1995; Nazroo, 1997; Chamba et al, 1999). Social class, for example, might explain many of the risk factors associated with conditions such as coronary heart disease, rather than ethnic origin per se (Bhopal et al, 2004). When looking at differences between how Latina and Anglo women in North America respond to infertility, Chavez (2004), concluded that ethnic differences were insignificant: age, education and marital status were more relevant. This is why we need to accept that in some ways minority ethnic populations may not be all that different from the general population. Similarly, ethnicity does not always equate with difference and disadvantage. Not every problem or difficulty a person encounters in attempting to gain access to appropriate health and social care can be attributed to his or her ethnic background. The challenge is to know when ethnicity makes a difference and mediates a person's experience and sense of identity and when it does not.

Institutional racism in the UK

Leaving aside discussions about ethnic identity, we have nonetheless to acknowledge that developing health and social care provision for ethnically diverse populations occurs against a backdrop of socio-economic disadvantage and social exclusion (Parekh, 2006). In straightforward empirical terms, ethnic minority populations tend to experience greater poverty than the general population. The average weekly net earnings of Bangladeshi men living in the UK are about 52 per cent below those of their white counterparts; after housing costs, 30 per cent of African-Caribbean families are classified as being on low incomes, compared to 20 per cent of the majority population (see Nazroo, 2006). Similar associations can be seen in other post-industrial societies (see Lopez, 2004). Given the link between poverty and ill health, it is perhaps not surprising that ethnic minority populations are also more likely to describe their health as poor, when compared to the majority ethnic population (Sproston and Mindell, 2006). Social exclusion of ethnic minority populations is not new (see Marshall, 1964), but continues to have an uneasy resonance in a multicultural society and in light of recent policy discussions on the dangers of segregation, estrangement and polarization (see Cantle, 2001, for specific discussion of UK policy and Kymlicka, 2001, for a more international perspective).

Institutional racism in health and social care

In the UK 'institutional racism' has become a popular explanation for public services' inability to respond to the needs of an ethnically diverse society (Karlsen, 2007, and see above). Institutional racism, however, is not a new concept. It was first introduced as an analytical category over 20 years ago (see Glasgow, 1980) and its political origins lie in the Black Power movements in North America of the 1960s (Zuberi, 2001). As such, it has analytical relevance beyond the UK. At its worst, straightforward, institutional racism occurs when the policies of an institution lead to discriminatory outcomes for minority ethnic populations, irrespective of the motives of individual employees of that institution. In effect, it is embedded in the assumptions informing organizational practices. Fertility services embody many aspects of institutionally racist practices. Minority ethnic families, for example, struggle to gain appropriate language support; have to engage with practitioners who make little concession to possible cultural differences in how the experience of fertility is interpreted; and are subject to ill-informed assumptions about their family relationships and reproductive practices (Schmid et al, 2004; Culley et al, 2006). Culley et al (2006), for example, specifically illustrate how practitioners struggle to address cultural norms and values as well as language needs of people of South Asian origin who experience fertility impairment.

The most obvious expression of institutional racism is where the needs of ethnic minority populations become ignored. Potential differences between those of an ethnic minority and the general population become disregarded because there is an underlying assumption that policies, procedures and practices are equally appropriate for everyone (see Karlsen, 2007). Service organizations have a historical tendency to adapt a so-called 'colour blind' approach which, by default, favours the majority white population, who come to represent the 'norm' around which service delivery becomes organized (Katz, 2002). Service users, for example, are assumed to have Western attitudes, priorities, expectations and values, to act according to Western ways, to speak English and understand the organization of public services (Parekh 2006). The inability of healthcare agencies to provide adequate support for those whose speak languages other than the dominant language is the most obvious example of this.

This example, however, demonstrates some of the subtleties in how we come to understand institutional racism. Language is not the only barrier to successful communication. In the UK 60 per cent of ethnic minority populations speak English as a first language. Difficulties in providing language support, although incredibly important, should be seen in perspective. There is an unfortunate history of conceptualizing language difficulties as one of the major problems facing public organizations as they struggle to engage with minority ethnic populations (Law, 2003). Even if there were sufficient resources for an expansion of provision of interpreters and greater translation of information into different languages, many fundamental problems would still remain. Barriers to communication are more than language specific and evoke cultural differences (Netto et al, 2001). Exploring this introduces another important theme in helping us understand institutional racism – the process by which the needs of ethnic minority populations become misrepresented.

If cultural diversity is recognized, it is often used against ethnic minority populations (Karlsen, 2007). Evidence suggests that healthcare problems of ethnic minority families can easily become associated with their supposed 'deviant' cultural practices and lifestyles (Ahmad and Bradby, 2007). As part of this process, the use of inappropriate myths and stereotypes, while purporting to explain the behaviour and beliefs of ethnic minority populations, do little more than corrupt their experiences, reducing them to essentialized, static entities (Atkin and Chattoo, 2007).

There are several ways in which this is done. Sometimes it is the consequence of ill-informed views about the cause of 'problems' presented by ethnic minority populations. Health and social agencies, for example, often identify 'ethnic' health and social 'problems' as arising from 'pathological' cultural practices (Mason, 2003). In the UK, there is a history of defining health problems faced by South Asian and African-Caribbean populations in terms of cultural deficits, where a shift towards a 'Western' lifestyle is offered as a solution to their problems. Policy examples include discussion on maternity and child health, diet and rickets (Law, 2003). Such views become embedded in the thinking of front-line practitioners working in health and social services; minority ethnic people become seen as 'a problem'. Young people suffering pain related to sickle cell disorders, for example, might be denied effective pain relief because of the mistaken perception that their ethnicity means they are more likely to get addicted to pain-killers than other patients (see Anionwu and Atkin, 2001). This occurs in both the UK and North America. The recent recognition of discriminatory treatment received by African-Caribbean and African-American men in the mental health services offers another sad reminder of a similar set of essentialist constructions of race and ethnicity (see Bhui et al, 2004). Further, 'poor childhood health' among Pakistani families often becomes attributed to consanguineous marriages, which carries an implicit (and misleading) criticism of cultural practices (see Ahmad et al, 2000, for a review of the evidence). However, other explanations – such as poverty, poor maternal health, inappropriate housing or inadequate service support – have more impact on childhood disability and illness than whether parents are married to a first cousin (see Bradford and District Infant Mortality Commission, 2006).

Ironically, certain problems associated with myths and stereotypes occur because authors want to be helpful and provide explanations that enable professionals to respond to the needs of a multi-ethnic society. Introductory notes on ethnic minority communities, present in training material for service practitioners, often follow this pattern. It is common, for example, to see one-page explanations of Muslim, Hindu and Sikh culture, to which patients' beliefs are expected to correspond. Such explanations tend to present static and one-dimensional views of cultural norms and values, devoid of context, allowing no room for individual interpretation (Atkin and Chattoo, 2007). They also create the illusion that they offer a solution to an extremely complex situation (Chattoo and Ahmad, 2004). Myths, often derived from simplistic accounts of Islamic beliefs, found in several text books, explain why prenatal diagnosis is sometimes withheld from Muslim families (Anionwu and Atkin, 2001). Termination is assumed unacceptable to such families because it is incompatible with religious values, although the empirical reality is far more complex, showing diversity in opinions (Tsianakas and Liamputtong, 2002). Termination is acceptable to some Muslim families but not to others. There is no single Islamic interpretation of the acceptability of termination and families' decision making processes are informed by their own values and beliefs, the views of other family members and the role of healthcare professionals (Atkin et al, 2008). The mistaken assumptions of practitioners, who struggle to accommodate such diversity, lead to a denial of opportunities for ethnic minority populations, pivotal to understanding institutional racism.

Evidence from the UK illustrates the conceptual benefits of institutional racism to the extent it provides us with a framework in which we can understand the general process whereby the needs and experiences of ethnic minority populations can either become ignored or misrepresented. Such benefits, however, need to be considered in relation to debates about the complex ways in which ethnic identity is expressed and becomes known to us.

The expression of ethnic identity in a multicultural society

Ethnic identity, as we have seen, involves active social construction within the context of a specific historical and political location. This is the dynamic space in which ethnic identity assumes meaning. But how does this find empirical expression in the UK and what lessons can be learned for other multicultural societies? Previous understandings, classifying people into ethnic groups according to their country of origin, seem no longer sustainable, particularly since nearly 40 per cent of those we regard as of an ethnic minority are born in the UK. Further, regional identification might emerge as important in explaining ethnic differences, even when people share the same country of origin. In Pakistan, cultural, linguistic and socio-economic differences occur between those who originate from the Punjab and those from the North West Frontier. Moreover, those who claim Mirpuri Punjabi heritage represent a cultural and linguistic community, distinct from the rest of the Punjab (Husband, 1996).

We also know that young people born and brought up in the UK have different experiences and expectations from their parents, who may still perceive themselves as 'immigrants'. Younger people of African-Caribbean origin, for example, are more likely to favour the label of 'Black' British, when compared to their parents and grandparents (Modood et al, 1997). A further expression of this process suggests young people are increasingly using religious affiliations, such as Muslim or Hindu or Sikh, in addition to ethnic origin, when describing their identity (Ahmad and Bradby, 2007), specifically reflecting changes in how ethnic identity becomes known to us, as we continually reassess our understanding of what constitutes a multicultural society (Kymlicka, 2001). I have already touched on this, although such processes embody a more fundamental point, reflecting how the complex and dynamic nature of ethnic identity is realized within the broader context of social change.

Some Muslim young people whose families originate from the Punjab, for example, criticize their parents' interpretation of Islam, as embodying cultural practices, which have nothing to do with faith (Atkin and Chattoo, 2007). Parents themselves express concerns that their children's religious values become corrupted by Western practices (see Ahmad et al, 2002). Indeed Muslim young people do seem to have a more flexible approach to Islam than their parents and emphasize the more personal aspects of their religious beliefs, known as personal prayer or *dua* as opposed to formal ritualized prayers or *salat* (Modood et al, 1997). Religion, however, is still an important way of life for Muslim young people, providing a moral structure, which has overall implications for their lives (Atkin et al, 2008). This broader process of social change also partially explains why Hindu and Sikh young people, although just as likely as their parents to acknowledge the importance of religious identity, are more likely to emphasize the importance of private spirituality (Modood et al, 1997).

There is considerable continuity in values between the different generations, as ethnicity and religious identity continue to reinforce each other. This is how social change becomes negotiated, although for ethnic minorities living in the UK this negotiation occurs within a broader political process, with which they must also engage, and which is – as we have seen – constructed within the context of discriminatory and essentializing discourses. Nonetheless, social policy and sociological as well as lay discourses often assume social change to be a consequence of acculturation or adoption of the values of the 'host society' by the younger generation of an 'immigrant' community. The perceived consequence is an inevitable conflict of values between the older and younger generation or, at worse, a generation of estranged young people, trapped between cultures (Dorling and Thomas, 2004). Research findings, however, do not support such a view and highlight the significance of negotiation within and between generations (Atkin and Chattoo, 2007, and see above).

Before leaving this discussion, there is one more general point I need to make: the complex and dynamic nature of ethnic identity requires public policy 'to recognize a wider range of cultural identities than it does at present' (The Parekh Report, 2000, p242). Recent changes in the process of immigration are beginning to change how we define ethnicity and imagine a multicultural state (see Robinson et al, 2007). Our sense of ethnic identity, therefore, is constantly shifting. Asylum seekers and refugees have become part of the political landscape and a feature of life in UK cities. Immigration from Central and Eastern Europe is adding to our sense of an ethnic white minority, which has never been a strong feature of UK debates about multiculturalism, despite the fact that some of these minorities have a long-standing history of settlement in the UK. The experience of disadvantage and discrimination for such groups is likely to be very different from the more established ethnic minority populations living in the UK. A recent headline in the *Yorkshire Post*, highlighting the growing cost of interpreting services for Yorkshire police forces, quoted a representative of the Taxpayers Alliance as saying that people who wished to live in the UK should be made to learn English (*Yorkshire Post*, Saturday 17 November 2008). The comments, however, were aimed at Polish and Russian migrants. This is an expression of how discussions about recent migrants have reinvented a discourse, first evident during the mass influx of migrants from the Caribbean and South Asia during the 1960s and 1970s, in which immigrants came to be seen as a threat, with the potential to undermine and swamp British culture (Craig, 2007). There is then the invisibility of several ethnic minority populations, such as those of Chinese origin, who by being evenly dispersed across the UK, do not have a large visible policy presence in any one region or locality. Similarly, it is easy to marginalize the experience of Travelling and Roma communities as well as those who regard themselves as being of Irish origin, in addition to the growing number of people who claim their ethnic origin as mixed heritage.

Improving outcomes

Returning to an issue raised in the initial preface, how can we translate what are at times extremely abstract and esoteric debates into improved outcomes for minority ethnic populations? True, a vast literature has emerged challenging the ways in which public care agencies engage with ethnic minority populations (Bhopal, 2007) and this has begun to influence service delivery in post-industrial societies (see Dominelli, 2004; Eisenbruch et al, 2004; Papadopoulos et al, 2004; Paez et al, 2008). Yet despite growing recognition of the value of culturally competent practice, practitioners still express uncertainty and a lack of confidence when discussing issues of ethnicity and cultural difference (Kai et al, 2007; Peckover and Chidlaw, 2007), often reverting to simplified generalizations, which they believe enables them to 'manage' ethnic diversity (Qureshi et al, 2000).

Describing and documenting disadvantage creates a potential jeopardy. Constantly highlighting the negative consequences of service provision can sometimes do little to advance thinking and practice (Levick, 1992): condemning everything while proposing nothing (see Bauman, 1992) and creating a sense of inertia, makes active engagement problematic. Policy and practice have not been particularly successful in translating evidence, outlining the process and outcomes of discrimination into tangible improvements in service delivery (Bhavnani et al, 2005). Offering an analysis of the problems facing minority ethnic populations is one thing; doing something about them is another.

Reflexive practice and communicative competence

An initial response to the existing gap between insight and practice lies in encouraging the role of the 'reflexive practitioner', who is able to respond to complex situations without relying on 'fact-files' or bureaucratized responses (see Evans and Harris, 2004), working with people's definitions of what is happening to them, rather than imposing one's own ideas, sanctioned through professional practice (Atkin and Chattoo, 2007). Such reflexivity could empower professionals to raise meaningful questions regarding the social context within which supportive relationships are legitimized and negotiated (see, for example, National Family and Parenting Institute, 2005) and begin to provide the cultural repertoire and confidence to engage with diversity and difference (Dominelli, 1988). This finds resonance in current debates about culturally competent practice in healthcare, which emphasizes the importance of getting practitioners to challenge their own values, develop understanding and sensitivity, and apply awareness and knowledge to appropriate practice (Papadopoulos et al, 2004).

The starting point for successful policy and practice guidance should, therefore, be an analysis of the present difficulties, an explanation of how practitioners currently make sense of these difficulties, and presentation of alternative ways of addressing the situation. Achieving this, however, is far from straightforward and requires a more fundamental change in professional culture and policy formulation. The focus should shift from 'understanding minority ethnic cultures' to 'how services respond to need'. Reflexivity is not simply about individual reflection on practice, but engagement with the ways in which institutional practices support the role of the 'reflexive' practitioner (see Alvesson, 2002). *Communicative competence*, or the process whereby practitioners – and the public agencies within which they work – frame, engage and legitimate the experiences of ethnic minority populations, becomes a fundamental part of this process (see Habermas, 1987). In effect, we need to understand the social, cultural and organizational context in which practitioners work. The reflexive practitioner cannot, of course, transcend the consequences of this context, in which different stakeholders attempt to define situations, speak with legitimacy and have others recognize the legitimacy of their views (see Bourdieu, 1977). He or she does, however, mediate its outcome by recognizing this role as an important part of reflexive engagement. Welfare provision in post-industrial societies is by its nature complex, full of ambiguities, contradictions, inconsistencies and compromises, which reflect a mix of individual and collective solutions.

Conclusion

Despite increasing awareness of the complex nature of culturally sensitive provision, health and social care agencies have been slow to provide accessible care, appropriate to the needs of minority ethnic populations. This is a feature of most post-industrial societies. Despite difficulties in making cross-country comparisons – where ideas about ethnicity can assume different meanings and have different outcomes, as a consequence of the specific historical, social, cultural, economic and political contexts in which they occur – several general themes can be identified. Various discriminatory practices – revealed in professional assumptions and organizational systems – reduce ethnic minority communities to monolithic, homogeneous wholes, defined by their ethnicity, religion or culture, with little reflection on differences within and similarities across ethnic groups, as marked by gender, religion or socio-economic position (see Culley and Hudson, Chapter 6). Structural issues of inequality emerge as a fundamental part of this process (see Gunaratnam, 1997) and this is the first theme – shaping how we come to know ethnicity. To this extent, institutional racism provides an ideal framework to make sense of the disadvantages faced by minority ethnic groups, as they try to gain access to appropriate healthcare. The discussion of institutional racism, however, also demonstrates the complexities evoked when trying to make sense of ethnic diversity. This understanding embodies 'being', and is where a person's sense of ethnic identity becomes meaningful. People emerge as active social agents who create and negotiate values and meanings within the context of their social identities, which in turn are realized within a broader social, political and historical context, which seeks to define ethnicity.

Finally, in facilitating successful policy and practice we have to remember that focusing on the needs of minority ethnic populations is not the same as responding to those needs. Despite a willingness on the part of the post-industrial states to tackle discrimination, public organizations not only struggle to reconcile key ideas – such as institutional racism and community cohesion – within a policy framework (Ratcliffe, 2004) but also often lack a political strategy to initiate change (Kymlicka, 2001). As Charles Taylor (1994) observes, to sustain multicultural societies, we need to develop 'the politics of difference' – in which there is a political commitment to ensuring diversity does not become the basis of inequality – rather than the 'politics of representation' – in which recognizing difference becomes confused with responding to it. There remains a longstanding and ongoing disparity between our understanding of the issues and our commitment to act. Overcoming this remains an ongoing challenge for policy and practice.

References

Ahmad, W. I. U. and Bradby, H. (2007) 'Locating ethnicity and health: Exploring concepts and contexts', *Sociology of Health and Illness*, vol 29, no 6, pp793–811

Ahmad, W. I. U., Atkin, K. and Chamba, R. (2000) 'Causing havoc among their children: Parental and professional perspectives on consanguinity and childhood disability', in W. I. U. Ahmad (ed) *Ethnicity, Disability and Chronic Illness*, Open University Press, Buckingham

Ahmad, W. I. U., Atkin, K. and Jones, L. (2002) 'Being deaf and being other things: South Asian deaf young people and identity negotiation', *Social Science and Medicine*, vol 55, no 10, pp1757–1769

Alvesson, M. (2002) *Postmodernism and Social Research*, Open University Press, Buckingham

Anderson, B. (2006) *Imagined Communities*, Verso, London

Anionwu, E. and Atkin, K. (2001) *The Politics of Sickle Cell and Thalassaemia*, Open University Press, Buckingham

Atkin, K. and Ahmad, W. I. U. (2000) 'Living with sickle cell disorder: How young people negotiate their care and treatment', in W. I. U. Ahmad (ed) *Ethnicity, Disability and Chronic Illness*, Open University Press, Buckingham

Atkin, K. and Chattoo, S. (2007) 'The dilemmas of providing welfare in an ethnically diverse state: Seeking reconciliation in the role of a "reflexive practitioner"', *Policy and Politics*, vol 35, no 3, pp379–395

Atkin, K., Ahmed, S., Green, J. and Hewison, J. (2008) 'Decision making and ante-natal screening for sickle cell and thalassaemia disorders: To what extent do faith and religious identity mediate choice?', *Current Sociology*, vol 56, no 1, pp77–98

Balen, R. and Crawshaw, M. (2006) *Sexuality and Fertility Issues in Ill Health and Disability*, Jessica Kingsley, London

Bauman, Z. (1992) *Intimations of Post Modernity*, Routledge, London

Bhavnani, R., Mirza, H. S. and Meetoo, V. (2005) *Tackling the Roots of Racism*, Policy Press, Bristol

Bhopal, R. (2007) *Ethnicity, Race and Health in Multicultural Societies*, Oxford University Press, Oxford

Bhopal, R., Hayes, L., White, M., Unwin, N., Harland, J., Ayis, S. and Alberti, G. (2004) 'Ethnicity and socio-economic inequalities in coronary heart disease, diabetes and risk factors in Europeans and South Asians', *Journal of Public Health Medicine*, vol 25, no 2, pp95–105

Bhui, K., McKenzie, K. and Gill, P. (2004) 'Delivering mental health service for a diverse society', *British Medical Journal*, vol 329, pp363–364

Bleich, E. (2003) *Race Politics in Britain and France*, Cambridge University Press, Cambridge

Bourdieu, P. (1977) *Outline of a Theory of Practice*, Cambridge University Press, Cambridge

Bradby, H. (2003) 'Describing ethnicity in health research', *Ethnicity and Health*, vol 8 no 1, pp5–14

Bradford and District Infant Mortality Commission (2006), *Summary Report*, BDIFV, Bradford, see www.bdimc.bradford.nhs.uk

Cameron, E., Badger, F., Evers, H. and Atkin, K. (1989) 'Black old women and health carers', in M. Jefferys (ed) *Growing Old in the Twentieth Century*, Routledge, London

Cantle, T. (2001) *Building a Safe, Just and Tolerant Society: A Report of an Independent Review Team*, Home Office, London

Chamba, R., Ahmad, W. I. U., Hirst, M., Lawton, D. and Beresford, B. (1999) *On the Edge: A National Survey of Minority Ethnic Parents Caring for a Severely Disabled Child*, Policy Press, Bristol

Chattoo, S. and Ahmad, W. I. U. (2004) 'The meaning of cancer: Illness, biography and social identity', in D. Kelleher and G. Leavey (eds) *Identity and Health*, London, Routledge

Chattoo, S. and Ahmad, W. I. U. (2008) 'The moral economy of selfhood and caring: Negotiating boundaries of personal care as embodied moral practice', *Sociology of Health and Illness*, vol 30, no 4, pp550–564

Chavez, L. R. (2004) 'A glass half empty: Latina reproduction and public discourse', *Human Organisation*, vol 63, no 2, pp173–188

Craig, G. (2007) 'Cunning, unprincipled, loathsome: The racist tail wags the welfare dog', *Journal of Social Policy*, vol 36, no 4, pp605–623

Culley, L. A., Hudson, N., Rapport, F. L., Katbamna, S. and Johnson, M. (2006) 'British South Asian communities and infertility services', *Human Fertility*, vol 9, no 1, pp37–45

Dominelli, L. (1988) *Anti-Racist Social Work*, Macmillan, London

Dominelli, L. (2004) *Social Work: Theory and Practice for a Changing Profession*, Polity Press, Cambridge

Dorling, D. and Thomas, B. (2004) *People and Places*, Policy Press, Bristol

Eisenbruch, M. M., Yeo, S., Meiser, B., Goldstein, D., Tucker, K. and Barlow-Stewart, K. (2004) 'Optimising clinical practice in cancer genetics with cultural competence: Lessons to be learned from ethnographic research with Chinese-Australians', *Social Science & Medicine*, vol 59, no 2, pp 235–248

Evans, T. and Harris, J. (2004) 'Street-bureaucracy, social work and the (exaggerated) death of discretion', *British Journal of Social Work*, vol 34, no 6, pp871–896

Giddens, A. (1991) *Modernity and Self Identity: Self and Society in Late Modern Age*, Polity Press, Cambridge

Glasgow, D. (1980) *The Black Underclass*, Jossey Bass, London

Gunaratnam, Y. (1997) 'Culture is not enough: A critique of multiculturalism in palliative care', in D. Field, J. Hockey and N. Small (eds) *Death, Gender and Ethnicity*, Routledge, London

Habermas, J. (1987) *Knowledge and Human Interests*, Polity Press, Cambridge

Hall, S. (1996) 'Introduction: Who needs identity', in S. Hall and P. du Gay (eds) *Questions of Cultural Identity*, Sage, London

Healy, J. and McKee, M. (2004) *Accessing Health Care: Responding to Diversity*, Oxford University Press, Oxford

Home Office (2004) *Strength in Diversity: Towards a Community Cohesion and Race Equality Strategy*, Home Office, London

Husband, C. (1996) 'Defining and containing diversity: Community, ethnicity and citizenship', in W. I. U. Ahmad and K. Atkin (eds) *Race and Community Care*, Open University Press, Buckingham

Jenkins, P. (1996) *Identity*, Routledge, London

Judd, D. (1996) *Empire: The British Imperial Experience*, Harper Collins, London

Kai, J., Beavan, J., Faull, C., Dodson, L., Gill, P. and Beighton, A. (2007) 'Professional uncertainty and disempowerment responding to ethnic diversity in health care', *PLoS Medicine*, vol 4, no 11, p323

Karlsen, S. (2007) *Ethnic Inequalities in Health: The Impact of Racism* (Better Health Briefing Paper 3), Race Equality Foundation, London

Katz, A. (2002) *Thwarted Dreams: Young Views from Bradford*, Young Voice Project, London

Kymlicka, W. (2001) *Politics in the Vernacular: Nationalism, Multiculturalism and Citizenship*, Oxford University Press, Oxford

Law, I. (2003) *Race, Ethnicity and Social Policy*, Prentice Press, Hemel Hempstead

Levick, P. (1992) 'The Janus face of community care legislation: An opportunity for radical opportunities?', *Critical Social Policy*, vol 12, no 1, pp75–92

Lopez, R. (2004) 'Income inequality and self-rated health in US metropolitan areas: A multi-level analysis', *Social Science and Medicine*, vol 59, no 12, pp2409–2419

Marshall, T. H. (1964) *Class, Citizenship and Social Development*, Cambridge University Press, Cambridge

Mason, D. (2003) *Explaining Ethnic Differences*, Policy Press, Bristol

Modood, T., Betthould, R., Lakey, J., Nazroo, J., Smith, J., Virdde, S. and Beishon, S. (1997) *Ethnic Minorities in Britain*, Policy Studies Institute, London

National Family and Parenting Institute (2005) *A Cultural Competence Toolkit for Working with Black, Minority Ethnic and Faith Families*, NFPI, London

National Institute for Clinical Excellence (2004) 'Fertility: Assessment and treatment for people with fertility problems', www.nice.org.uk/nicemedia/pdf/CG011fullguideline.pdf, accessed 20 June 2008

Nazroo J. (1997) *The Health of Britain's Ethnic Minorities*, Policy Studies Institute, London

Nazroo, J. (2006) 'Demography of multicultural Britain', in J. Nazroo (ed) *Health and*

Social Research in Multiethnic Societies, Routledge, London

Netto, G., Gaag, S., Thanki, M., Bondi, L. and Munro, M. (2001) *A Suitable Space: Improving Counselling Services for Asian People*, Policy Press, Bristol

Nizalova, O. (2000) 'Economic and social consequences of maternal protection: A cross-country analysis 2000', www.gdnet.org/pdf/948, accessed 20 June 2008

Ouseley, H. (2001) *Community Pride not Prejudice: Making Diversity Work in Bradford*, Bradford Vision, Bradford

Paez, P. A., Allen, J. K, Carson, K. A., and Cooper, L. A. (2008) 'Provider and clinic cultural competence in a primary care setting', *Social Science & Medicine*, vol 66, no 5, pp1204–1216

Papadopoulos, I., Tilki, M. and Lees, S. (2004) 'Promoting cultural competence in health care through a research-based intervention in the UK', *Diversity in Health and Social Care*, vol 1, no 2, pp107–116

Papastergiadis, N. (1982) *Dialogues in the Diasporas: Essays and Conversations on Cultural Identity*, Rivers Oram Press, London

Parekh, B. (2006) *Rethinking Multiculturalism: Cultural Diversity and Political Theory*, Palgrave Macmillan, Basingstoke

Parekh Report, The (2002) *The Future of Multi-Ethnic Britain*, Profile Books, London

Peckover, S. and Chidlaw, G. (2007) 'The (un)certainties of district nurses in the context of cultural diversity', *Journal of Advanced Nursing*, vol 58, no 4, pp377–385

Qureshi, T., Berridge, D. and Wenman, H. (2000) *Where to Turn? Family Support for South Asian Communities: A Case Study*, Joseph Rowntree Foundation, York

Ratcliffe, P. (2004) *'Race', Ethnicity and Difference*, Open University Press, Maidenhead

Robinson, D., Reeve, K. and Casey, R. (2007) *The Housing Pathways of New Immigrants*, Joseph Rowntree Foundation, York

Royal Society of Arts (2006) *Migration: A Welcome Opportunity*, Royal Society of Arts, London

Schmid, J., Kirchengast, S., Vytiska-Binsorfer, E. and Huber, J. (2004) 'Infertility caused by PCOS: Health related quality of life among Austrian and Moslem immigrant woman in Austria', *Human Reproduction*, vol 19, no 10, pp2251–2257

Smaje, C. (1995) *Health, 'Race' and Ethnicity: Making Sense of the Evidence*, King's Fund Institute, London

Sproston, K. and Mindell, J. (2006) *Health Survey for England: The Health of Minority Ethnic Groups*, The Information Centre, London

Taylor, C. (1994) *Multiculturalism: Examining the Politics of Recognition*, Princeton University Press, Princeton, NJ

Tsianakas, V. and Liamputtong, P. (2002) 'Prenatal testing: The perceptions and experiences of Muslim women in Australia', *Society for Reproductive and Infant Psychology*, vol 20, no 1, pp232–241

Williams, F. (1996) 'Race, welfare and community care, a historical perspective', in W. I. U. Ahmad and K. Atkin (eds) *Race and Community Care*, Open University Press, Buckingham

Zuberi, T. (2001) *Thicker than Blood: How Racial Statistics Lie*, University of Minnesota Press, Minneapolis, MN

4

Representation of Ethnic Minorities in Research: Necessity, Opportunity and Adverse Effects

Mark Johnson and Theda Borde

Introduction

With an international growth in migration and increasing socio-cultural diversity in their populations, most European countries are faced with a need to look at healthcare services, asking how far these are equipped and trained to meet the differing needs of diverse groups. Subsequently staff will need to be supplied with the knowledge and skills required to develop appropriate healthcare and health promotion for all. This will require access to a good evidence base of research on ethnic difference and/or the impact of migration on health in general. While true overall, this is of particular relevance to life activities and aspects of healthcare that deal with intimate and culturally constructed aspects of society, such as family creation. These involve issues of culture that go far beyond 'race', migrant status and language. Therefore, we argue, research in the field of reproductive technologies in particular should be looked at in the broader context of a concept that we term 'diversity-health'. This approach takes into account migration, culture and ethnicity as well as classical socio-demographic factors such as education, social status, age and gender.

Critical issues in the process of change to reflect the new reality of a diverse population include four basic stages. First, it is necessary to establish that there are inequalities (if there are) and that change is necessary – and for some national settings, this case is still being made. However, the research reported in this volume leads us to feel confident that this stage has been reached in relation to the reproductive health of migrant minorities in Europe. Second, we need ways of understanding the processes that create inequalities. Third, it is essential to have ways of measuring change in outcomes or satisfaction in order to present evidence of improvement. Finally, in an evidence-based policy world, as research is only really of value if it supports practice, we need robust evaluations of interventions that may work to reduce inequalities. At present there are very few of the last, since the second and third stages of the journey have yet to be achieved in most cases.

In this task, we face problems in the comparability of cross-national data and are hindered by a lack of research methods and instruments that adequately include and represent diversity in its complexity. This includes, in particular, the need to capture and measure the diversity present in the migrant population, and ways of conducting meaningful research in and with these populations, taking account of differences in language and literacy between groups. At present, the healthcare needs of migrants and ethnic minorities are scarcely understood and only infrequently met in most European countries, because although migrants and ethnic minorities may make up a large percentage of the population, they are still considered to be 'hard to reach' and 'difficult to interact with'. These assumptions certainly represent the general ethnocentric (or even Eurocentric) view across Europe and require addressing by cross-national discussion, cooperation and action to enlarge the evidence base on multicultural health concerns.

We describe here some of the methodological challenges of including migrants and ethnic minorities in research. These are illustrated by examples from studies in Germany considering German and Turkish migrant patients, and from the UK where the white majority and South Asian minority communities have been involved. These reveal the complexity of the subject, highlighting the role of culture, language and literacy as well as the necessity of studying these groups through specific enquiries. They also show the possible adverse effects and positive opportunities that present themselves in the representation of diversity, ethnicity and migration, and the implications of adding a further layer of complexity to health research.

The analysis of inequalities in health has historically been informed by the debate on class or socio-economic inequalities in health, sometimes to the exclusion of consideration of ethno-cultural difference – or alternatively, has been dominated by a 'culturalist', 'blame the victim' approach (see Ahmad et al, 2000). It is important to recognize that migrants and ethnic minorities in most countries occupy the most deprived socio-economic positions and that 'class' has a role to play. Health inequalities can result from broader societal health determinants: socio-economic factors and psychosocial stress have a direct impact on health behaviour, morbidity and mortality. However, for migrants and ethnic minorities *additional* factors may impact on access to healthcare as well as on the outcome of the healthcare they receive. In relation to reproductive health, the process of migration, disturbed family patterns, changing cultural milieus and breaking social linkages may also be significant.

Culture has an impact on gender roles, the value of children, their preferred sex, the meaning of infertility and influences coping with infertility. There are also cultural and ethical influences on attitudes towards reproductive technologies (Culley et al, 2006; van Balen, Chapter 2). Religion may be a part of this, or may overlap and add to linguistic or 'geographical' cultures, so that Muslims may have very different views from Sikh or Hindu people whose roots are in the same region of India, as is the case for the British Asian population, many of whose roots lie in the Punjab region, but whose language and religions differ widely.

We also argue that the lack of a reliable research evidence base is itself a form of 'structural' or institutional racism (Macpherson, 1999; Atkin, Chapter 3), and we therefore concentrate on two aspects of organizational failure that contribute to this: the need to capture or record the 'ethnic' origins of people (both in specific research activity and in monitoring which permits and supports epidemiological research), and second, the need to obtain 'insider' accounts of health beliefs and behaviours belonging to members of those groups, so that any observed differences may be attributed to meaningful differences in behaviour or experience. Under this heading, we shall demonstrate that the need to work across boundaries of language, and to accommodate a lack of literacy in written language, are crucial elements in research. To do that, we draw on research studies in which we have been involved that illustrate these particular factors. Two of these focused on research methodology issues: data collection among South Asians from Bangladesh and Pakistan/Kashmir in Birmingham (Lloyd et al, 2008), and information understanding with Turkish and German women in Berlin (Pette et al, 2004). The other two concerned reproductive healthcare: infertility among South Asian people in Britain (Culley et al, 2006), and inpatient gynaecology in Turkish and German migrant women in Berlin (Borde et al, 2002).

Recording diversity: Availability and comparability of cross-national data on 'ethnicity'

Research cannot be totally distinguished from service delivery, since for epidemiological studies we are usually reliant on routine monitoring data, and this also provides the 'baseline' against which change can be measured, and validates the categories used in specific studies. However, in most European countries neither migration-relevant information nor ethnicity is routinely coded in administrative health data. The lack of data often derives from problems in representing 'diversity' in both official demographic statistics and research. There are problems in finding indicators that describe differences relevant for health-related issues, as well as national sensitivities in recording such data (Johnson, 2001). In international studies we find various terms and definitions used to differentiate ethnic minorities and/or migrant populations; for example: migrants, migrant background (of parents), 'foreigners', 'born in', Turks, Bangladeshi and so on. Even within a country the terms and definitions used are frequently used inconsistently (Bhopal and Donaldson, 1998). Indeed, even in the US, where the 'race' question and legislation since the 1964 Civil Rights Act have led to the extensive and detailed use of ethnicity/racial monitoring in research, the question has not been fully resolved nor consistently applied (Lin and Kelsey, 2000).

European anti-discrimination guidelines following the 'Amsterdam Treaty' and its 'Section 13' (European Union, 1997) include ethnicity and race as discriminatory factors and demand statistical data addressing migration and socio-cultural diversity in the population, but these have yet to be routinely implemented. In every national setting, different underlying concepts are used, or different labels used for similar groups – while at the same time, some terms (such as 'Asian') may mean different things in different places. Differing definitions and categories representing ethnicity or migration background result from historical and political debates which are also reflected in the systems of representation of ethnic minorities. Thus, in the US, key elements of classification are 'race' (skin colour – black and white), linguistic (Hispanic/Non-Hispanic), 'aboriginality' (Native American/other) and the geographical region of origin of migrants (ancestors). Countries of migrant settlement, like Canada and Australia, focus on migrant/aboriginal origins and language, while many European countries differentiate in statistics mainly between 'natives' and 'foreigners' based on the concept of citizenship or birthplace and migration. Britain follows the former models, Germany the latter.

While race/ethnicity concepts include several markers that have historically led to exclusion and discrimination, differentiation by citizenship alone does not supply the research information needed to improve healthcare, because ethnic minorities and (descendants of) migrants may obtain the citizenship of the country they live in by naturalization or by birth. Equally, birthplace alone may be misleading, especially for former colonial powers where members of families who served in territorial administration or missionary and similar work may well have children born 'overseas'. At one point in 1970s Britain, it was suggested that between 5 and 10 per cent of the 'New Commonwealth-born' population was 'white British', of such origins. Therefore, the opportunity to represent ethnic minorities in their heterogeneity should be grasped in both routine data collection and investigations like patient surveys as well as in intervention studies. Additional to standard indicators on socio-demographic aspects (age, sex, education, socio-economic status, etc.), any questionnaire used in surveys or routine descriptive data on service users should systematically include variables that allow comparative analysis and to evaluate the impact of migration, ethnicity and integration. These are at least:

- citizenship
- place of birth
- duration of residence in the country
- primary language
- self-defined ethnicity
- self-estimated knowledge of the official language
- religion.

On the other hand, the *place of birth of parents*, which is a proxy indicator often collected for migration-based analysis, will be redundant if the variables named above are included.

The situation in Germany

In the Federal Republic of Germany the population of foreign citizenship rose steadily from 1951 to 2005, reaching approximately 7.2 million people in the late 1990s (according to the foreign resident registry of the Federal Statistical Office). Problems of interpreting these data multiplied after changes in the Naturalization Law in 2000, allowing easier access to German citizenship for immigrants and their children. As a result of those regulations the level of births of non-German children in Germany decreased from 12.4 per cent in 1999 to 6 per cent in 2000. Additionally, approximately 4.4 million migrants of German background ('Aussiedler'), who have come to Germany since 1950 as refugees or late repatriates and obtained German citizenship on their arrival in Germany, are not represented as migrants in these statistics, although they share in the objective experience of migration. Among these are 2.2 million people who emigrated from the former Soviet Union, many of whom are quite culturally distinct. Thus demographic and research data based on citizenship has become problematic because the category no longer reflects the real implications of migration and cultural diversity since changes in immigration law (*Zuwanderungsgesetz*, 2005) initiated a shift in German policy from denial to acceptance of being an immigration country. This was followed not only by a discourse on the necessity of new data collection systems representing ethnic variety and the migration background of the population, but also the micro census held in 2005 (Federal Statistical Office, 2006). According to this representative survey, every fifth citizen of the Federal Republic of Germany has a migrant background (based on own and both parents' place of birth outside of the country) – this being a broader definition of a migrant, beyond the criterion of 'foreign citizenship' hitherto used. In the current report of the Federal Statistical Office on 'Births in Germany' it is still not possible to distinguish between naturalized and native German mothers. Of the 673,000 children born in Germany in 2006, 96 per cent had a German and 4 per cent a foreign citizenship. In 52 per cent of the German children, both parents had German citizenship and in 6 per cent both parents had a foreign nationality; in the remaining 42 per cent either the mother or the father was German.

This 'multicultural reality' has certainly reached medical practices and obstetric clinics in urban and industrial areas in Germany, where the proportion of migrant women among women giving birth may make up to 40 per cent – as is the case in the Charité Virchow Clinic in Berlin. Since data are not regularly collected on either migration background or ethnicity in health services, in our study on the utilization of emergency departments by migrants and German patients in Berlin (Borde et al, 2003; David et al, 2006), we asked patients for information about their migration background. Within the study sample of 815 patients (aged 15–65 years) we found that the percentage of 'non-Germans' (versus 'Germans') varied from 38 to 55.7 per cent depending on the definition this category was based on. Since in the context in Germany described above neither the category 'citizenship' (38 per cent) nor 'country of birth outside of Germany' (42.2 per cent) is appropriate to indicate ethnicity, we referred to the 'self-defined primary language' (55.7 per cent) for comparison purposes in that study, and this has proved to be a useful proxy indicator of diversity for other research.

The situation in the UK

For many years in Britain there was considerable resistance to the idea of recognizing ethnic difference in official statistics or research, especially in relation to healthcare and health outcomes (Johnson, 1998). Indeed, it was often argued that healthcare professionals would be morally incapable of discriminating on grounds of race. Further, the case was made that asking for, or recording 'ethnic origin' data was in itself potentially discriminatory, while at the same time recognizing that the absence of records meant that it could not be shown that discrimination was happening (Drew, 1980). There were also heated debates about precisely what was to be measured, and whether 'ethnic group' was a legitimate or intellectually coherent category (Royal Statistical Society, 1983; Burchard et al, 2003). Eventually, after a set-back in 1981 when the question was dropped at the last moment from the UK Census, the Office for Population Censuses and Survey incorporated a composite question into the 1991 census. It then became possible to refer to a robust denominator, and to use survey data to make estimates of national and local levels of need (Bardsley and Lowdell, 1999). Since that time, the political and bureaucratic consensus has changed, and current guidance from the Department of Health notes explicitly that health and social care services provided by the National Health Service (NHS) and local councils should treat their patients, service users and staff with fairness, dignity and respect. They must aim to deliver appropriate, personalized services to the diverse communities they serve and to be employers of choice that recruit, develop and retain the best talent from all communities. The Department and NHS have therefore published a *Practical Guide to Ethnic Monitoring* (Department of Health, 2005) in order to assist NHS and social care agencies to monitor performance with regard to race equality, and to compile, review and actively use Race Equality Schemes (required under the Race Relations Amendment Act 2000) to effect necessary improvements. These data – and the categories embodied in them – can then be used in research and as a basis for comparison, as well as in recommendations arising from research (for example, into the risk of developing heart disease: Hippisley-Cox et al, 2008).

The pragmatic, compromise categorization finally adopted for the UK 2001 Census, and now widely used in UK research, combines elements of 'race' and geographical origin. Four basic high-level categories are offered – White, Black, Asian, Chinese/Other, and one for 'Mixed' origins. Within each of these, people are asked to choose one of a number of sub-categories – or to write in a label of their own choice and select 'Other'. For white people, the options of British, Irish and Other are offered; for Black, African and Caribbean; while for Asians the choice is between the national origins of Indian, Pakistani and Bangladeshi. There is also a separate question on religious identity. A new proposed question on 'national identity' has been suggested for the next Census in 2011, as well as one on 'what languages can you understand, speak or write', which might be of value in research and planning services – and resembles the question used by German researchers. The main change which is likely to be adopted is the addition of two new 'other ethnic group' sub-categories: Arab and Gypsy/Romany/

Traveller, which will meet some needs but will still not allow identification of the currently growing groups of Somali and Eastern European origin.

Such research or 'epidemiological' information can of course also show that a service such as physiotherapy or occupational therapy (Johnson et al, 1996) or coronary revascularization (Fitzpatrick et al, 2005) is not actually discriminating, but may be operating in a different way for specific groups of clients. Similarly, the data can be used, as they have been in Liverpool (Public Health Sector Group, 2000), to develop services and ensure that developments are targeted to meet the real needs of patients and service users.

Language, literacy, education and inclusion of minority groups in research

When a minority group has been identified as distinctive, it is necessary to undertake research with its members to locate the causes of difference and inequality. These may be located within the structures of the majority society, or derive from cultural values and practices or understanding of the minority group. Only by research can this be resolved. However, it must not be assumed that the research methodology used in epidemiological and other studies of the majority community can be used unchanged with migrants and their descendants. Not only will language differ, but also concepts may be culturally distinct, and literacy may render standard survey tools and diagnostic instruments of little value (Collins and Johnson, 2007). A few examples may serve to illustrate some of the complexities involved, but also show the value of examining these to highlight and address issues arising in practice. Further, they may also draw attention to hidden cleavages in the 'native' or dominant majority population, since 'white' is also a complex community and not a homogeneous category (Tutton, 2007). We also show, by reference to our own research, some of the means, including additional questions and the use of qualitative methods, which help to overcome the problems identified in these examples.

Language and culture in gynaecology: Two German examples

Two studies from Berlin on the impact of migration and ethnicity on the health-care situation exemplify central factors that need to be measured, understood and considered in health research. These include, as a minimum, migration status (age at time of immigration, period of residence), proficiency in the dominant language, social situation (integration in working life, economic status), level of education (reading and writing abilities, health literacy), and of course cultural aspects (ethnicity, religion, values, norms) and gender.

A comparative study on the quality of care for Turkish migrant and German women in a gynaecological ward (Borde et al, 2002) showed that in the migrant sample originating from Turkey (n=262) 32 per cent of the women (age 15–70 years) belonged to the first generation of migrants who came to Germany mostly in the 1980s, 43 per cent were second generation (born or grew up in

Germany), and 19 per cent migrated after marriage to a '2nd generation' husband, while 6 per cent had come as asylum seekers. Related to the time of residence and the circumstances of migration and integration, 37 per cent of the sample were found to have only little knowledge of the German language, whereas 30 per cent estimated their knowledge to be fairly good and another 33 per cent claimed a good proficiency in German.

In order to include migrant women with little knowledge of the German language and with low literacy skills in the patient survey, we adapted the data acquisition process to the requirements of this sample and offered question-naires in German and in Turkish as well as oral standardized interviews using the questionnaire. Only 39 per cent of the migrant women chose the German version of the questionnaire whereas the rest preferred to read and fill it out in Turkish. While 68 per cent of the Turkish sample were able to fill out the ques-tionnaire by themselves, we found that about one-third of these women could only be included in the study by utilizing standardized oral interviews, due to their partial or total illiteracy. One in eight (12.4 per cent) of the 'Turkish' patients reported that Kurdish was their primary language. This aspect of access needs to be considered, especially in migrant groups of higher age and in migrant women who may not have had a basic school education. The data on educational attainment in the migrant women of this study showed that 10 per cent had never attended a school and 40 per cent had only completed five years of elementary school in Turkey. Secondary school-leaving qualifications had been reached by 38 per cent of the migrant and 53 per cent of the German women, while an A-level had been accomplished by 12 per cent (Turkish) compared to 42 per cent (German).

Basic knowledge acquired in school as well as access to other qualified sources of information had a direct impact on the basic health knowledge. When asked to match the terms 'ovary', 'fallopian tube', 'uterus', 'os uteri' and 'vagina' to an illustration of the female genitalia, the proportion of Turkish women with no answer was approximately 30 per cent, while 10 per cent of German women were likewise unable to answer. (This also suggests that even among the 'major-ity' population there may be problems with research instruments.) In addition, the frequency of correct matches also demonstrated striking differences between the two groups (Table 4.1) (David et al, 2000).

Table 4.1 *Knowledge about female genitalia: Matching up the names of female genitalia to an illustration*

	Correct answer		Incorrect answer		Not known	
	German	Turkish	German	Turkish	German	Turkish
Fallopian tube	77.5	55.5	13.1	11.1	9.4	37.4
Ovary	78.8	59.5	8.8	8.0	12.5	32.4
Uterus	84.1	63.0	1.9	3.8	14.1	33.2
Cervix	85.3	60.3	1.9	4.6	12.8	35.1
Vagina	86.9	63.0	1.3	5.7	11.9	31.3

Note: German women n=320; Turkish women n=262.

The results of a study on German and Turkish women's understanding of information or informed consent (Pette et al, 2004) show the impact of providing information that does not consider different requirements for differing patients. When comparing the concordance of patients' knowledge on their diagnosis and the therapy they received in the gynaecological ward with the data from the patients' files, we found that the Turkish migrant women were less informed compared to the German sample. More striking is the fact that while a small gain in correct answers was observable for the German sample from the day of admission to the day before discharge from the ward, among migrant women in the clinic a significant loss of information became visible. This effect was mainly observable for migrant women with little knowledge of the German language and may be explained by the fact that professional interpretation of the doctor–patient communication was not provided and the use of lay interpreters (mostly relatives) reduced the quality of the information given. The fact that even migrant women with good German language proficiency did not reach as high scores as the German sample not only calls for the implementation of interpreting services but requires a closer look at health literacy and the promotion of empowerment to enable participation and shared decision making among migrant women.

Problems of literacy and education: Two UK studies

It is well established that the health status of minority ethnic groups in the UK is often compromised, and this is particularly the case in relation to Type 2 diabetes, where the prevalence is at least five times greater in South Asians living in the UK compared to white Europeans (Erens et al, 2001). This is a chronic condition that requires self-management on a daily basis, along with support from a range of healthcare professionals (and in this respect has similarities with infertility). There are specific concerns with regard to supporting the appropriate self-management of all these conditions in minority ethnic groups, due to particular cultural and communication difficulties, but studies to improve both health service delivery and self-care may be compromised because established ways of collecting data are often inappropriate for these groups (Hunt and Bhopal, 2003; Collins and Johnson, 2007). High levels of illiteracy and lack of fluency in English are two key issues. Furthermore, in particular South Asian groups in the UK there is no agreed written form of the main spoken language; two such groups being those from the Sylhet region of Bangladesh, whose main language is Sylheti, and those from the Kashmir region of Pakistan, whose main language is Mirpuri. Both these groups are well represented (more than 50 per cent of patients) in the diabetes outpatient department of Birmingham Heartlands Hospital and have been the focus of a recent intervention study in diabetes complications risk-factor reduction (O'Hare et al, 2004). However, as Bhutta (2004) states, the notion of illiteracy does not mean that potential participants are unable to comprehend complex information, but it does mean that information may need to be presented in alternative ways.

A study was therefore designed to develop new ways of collecting data in South Asians with diabetes (Lloyd et al, 2008). An integral part of the study was to develop standard ways of obtaining informed consent, from people from minority ethnic backgrounds who were non-literate, which could be applied to other minority ethnic groups involved in research. The development of a standardized process of obtaining consent in situations such as these could lead to greater involvement of non-literate people in research in general. The research was to be based on the qualitative methodology known as 'focus group' discussion. The focus group sessions were gender and language specific, and were facilitated by one of two researchers along with either a lay member of the community (Sylheti groups) or an Asian link worker (Mirpuri groups). A sensitive assessment of literacy skills in both English and Bengali/Urdu was made by the researchers during recruitment (see Table 4.2). This was developed by the researchers in consultation with the Asian link workers, based on their knowledge of the study populations involved.

Table 4.2 *Screening questions pertaining to ethnicity and language/literacy*

What is/was your country of origin/ethnic origin?	☐ Bangladesh	☐ Pakistan
	☐ Others (non-eligible)...............	
Which part of Bangladesh/Pakistan are/were you originally from?	☐ Sylhet	☐ Dhaka
	☐ Punjab	☐ Mirpur
	☐ Other...............................	
Can you speak in English?	☐ Yes	☐ No
Can you read and write in English?	☐ Yes	☐ No
Can you speak in Bengali/Urdu?	☐ Yes	☐ No
Can you read and write in Bengali/Urdu?	☐ Yes	☐ No
What language/dialect do you usually use at home?	☐ Bengali	☐ Sylheti
	☐ Urdu	☐ Mirpuri
	☐ Others...............................	
What language/dialect do you usually use to speak with other people (friends, other members of same community/different community)	☐ Bengali	☐ Sylheti
	☐ Urdu	☐ Mirpuri
	☐ Others...............................	

The screening questionnaire was developed to ensure sensitivity when asking about literacy skills. No further difficulties were encountered in the use of the screening tool and so no further adaptations were required. The research highlighted some of the difficulties as well as the possibilities of conducting research with members of minority ethnic communities. In particular we found it more difficult to recruit and sustain participation in those individuals who spoke Mirpuri. Ways of maintaining interest and motivation still require developing, even when research assistants are able to engage with research participants through a common language. A previous report has also highlighted difficulties in recruiting individuals from minority ethnic communities and suggests that a range of methods may be necessary in order to maximize response rates (McLean and Campbell, 2003; Lloyd et al, 2006).

In research into infertility, the use of consensus or 'focus group' discussions has become quite common. However, few of these enquiries have addressed the specific views of minority ethnic groups, partly because of the apparent unwillingness of such groups to participate in conventional forms of research (Culley et al, 2007). All that is really required, however, is a small degree of flexibility and a willingness to accept that minority communities have rights and strengths. It may be necessary to make 'culturally appropriate adaptations' to conventional focus group techniques, such as the need to have single-gender groups with gender-matched facilitators, and to restrict their membership to an age-cohort to avoid problems of deference or concealment across generations (Winslow et al, 2002; Johnson, 2006). Training will almost certainly be needed for community-based facilitators who are locally trusted and speak relevant dialects, and their views may need to be heard in relation to the phrasing or use of acceptable terms and prompts. There are other bonuses from this, however. Focus groups are accessible to those who cannot read or write, and may provide safe space for people to discuss sensitive and personal issues such as racism and bullying that are rarely revealed to strangers or in formal settings, especially when these are seen as 'community' rather than individual detractions. In our study, we also found, however, that a more 'feminist' and less conventional approach was required, including sharing of both power and information. It is very hard to discuss or give an opinion on a technique on which one has little or no experience or information: by training facilitators to explain some of the technologies involved in fertility treatments, we were able to tap into a rich vein of discussion and reaction that would otherwise have remained silenced. Because they were trusted members of the community, their explanations were valued and accepted (and explained in culturally sensitive ways) while the lack of deference to 'expertise' normally extended to clinical staff meant that honest opinions were expressed. In order to do this, however, we had to create new educational materials, which in the end left a lasting legacy and benefit to the communities 'researched on' (Culley et al, 2006).

Conclusions and some suggested solutions

In summary, to address inequalities in health arising from ethnicity or migrant origin, it is not only necessary, but also possible, to record additional data that describe important demographic issues such as 'ethnicity', language, religion and migrant status, which will have a bearing on healthcare delivery and its planning. Further, additional research will be needed, adopting as far as possible an 'insider' perspective, taking account of language and cultural differences, to untangle the specific needs and understandings of the migrant/minority groups and the barriers that might exist to equality of access and enjoyment of services.

Some of the studies reviewed in this chapter have shown that a combination of social, educational and migration factors and the lack of use of healthcare concepts tailored to the specific needs of a diverse population lead to disparities in the access and outcome of information communication and healthcare. Language is important in transmitting – and accessing – ideas, but culture also plays an important role in defining values, beliefs about health and subjective theories of illness or the role of people involved in health as well as expectations of caregivers and treatment. By viewing issues through a 'culturally focused' lens, additional insights may be derived. For example, new categories of explanation in relation to reproductive health were yielded for migrant women with Turkish and Kurdish backgrounds which did not occur in the comparative German sample and seemed to be linked to their specific cultural backgrounds. These were 'many pregnancies and births', 'microbes and lack of hygiene', 'coldness', 'lack of knowledge' as well as 'magic influence' (in two cases for miscarriage) (David and Borde, 2001). Culture plays an important role concerning attitudes towards reproductive technologies. For example, Turkish couples (compared to German couples) are more willing to start in vitro fertilization (IVF) treatment at a younger age, their intensity of 'wish for a child' is significantly stronger; they are less likely to set time limits for IVF treatment and much less likely to see adoption as an alternative in case of infertility (Yüksel et al, 1996; Gacinski et al, 2002; Kentenich, 2002).

Engaging with minority ethnic communities in research, as these studies show, can be a positive experience, but is not without its difficulties. Different perceptions of skills, knowledge or expertise (and power) may exist on the part of professional groups and lay members of the community and can be problematic. Johnson has argued elsewhere (Johnson, 2006) that most research involves lay people only in terms of their use as fieldworkers or research staff, employed because of the ability to speak a particular language or gain access to certain groups. Research can only be improved by adopting some of the techniques and approaches necessary when working with minority and migrant populations. Thus, such research typically requires active attempts to engage with minority ethnic communities, including open discussion between the research participants and all the members of the investigative team. Studies should incorporate time for reflection and discussion between investigators, research assistants and service providers, which will facilitate the sharing of information and the taking on board of alternative perspectives with regard to the research process, the content of the questionnaires, participant information sheets and so on. Research assistants might not seek to be detached from the world of the minority/migrant, but to immerse themselves in the activities of the clinicians, shadowing Link Workers (minority ethnic liaison workers – similar roles are now found in many European hospitals: see the Migrant Friendly Hospitals project: www.mfh-eu.net/public/home.htm), so that they become accustomed to the patient experience and gain insights into care from a service-user perspective. and ensure that the findings of research are not only 'scientifically valid' but also of value to service users or the 'researched people'.

In conclusion, we have sought to build on the work of Maschewsky-Schneider and Fuchs (2000) and earlier researchers, who were concerned to avoid gender bias, and propose the following methodological and theoretical steps in the research process to ensure adequate representation of ethnic minorities. These should not only be applied in specific studies of ethnic minorities or migrants but should be considered in any investigation in health research and studies evaluating the quality of healthcare. Questions to be considered stepwise in the research process are:

1 *Research question:*
   Are ethnic/migrant minorities explicitly included in the research question? If not, why?
2 *Theoretical background:*
   Are differences known to relate to various ethnicities adequately considered in theory? Has the impact of migration and/or cultural change been addressed?
3 *Literature review:*
   Does the literature reviewed include research results from the different countries of origin of ethnic/migrant minorities concerned?
4 *Study design and sample:*
   Are ethnic/migrant minorities adequately represented in the sample? How are they identified or located?
5 *Instruments:*
   Are the research instruments appropriate to represent differences for ethnic minorities?
   Are the research instruments suited to reflect the perspectives or understanding and literacy of ethnic/migrant minorities?
6 *Data collection:*
   Do the methods for data collection allow adequate access to all ethnicities?
7 *Data analysis:*
   Are the statistical methods suited to explore ethnicity specific aspects?
8 *Data presentation:*
   Are the results presented for different ethnicities or migrant groups – and within these according to potentially relevant factors of religion, age, education, etc.?
9 *Conclusions and recommendations:*
   Do the conclusions and recommendations reflect differences in the life circumstances of ethnic minorities?

References

Ahmad, W. I. U., Atkin, K. and Chamba, R. (2000) '"Causing havoc among their children": Parental and professional perspectives on consanguinity and childhood disability', in W. I. U. Ahmad (ed) *Ethnicity, Disability and Chronic Illness*, Open University Press, Buckingham

Bardsley, M. and Lowdell, C. (1999) *Health Monitoring for Black and Minority Ethnic Groups: A Discussion Paper*, The Health of Londoners Project, London

Bhopal, R. and Donaldson, L. (1998) 'White, European, Western, Caucasian, or what? Inappropriate labeling in research on race, ethnicity, and health', *American Journal of Public Health*, vol 88, no 9, pp1303–1307

Bhutta, Z. A. (2004) 'Beyond informed consent', *Bulletin of the World Health Organization*, vol 82, pp771–777

Borde, T. (2002) 'Patientinnenorientierung im Kontext der soziokulturellen Vielfalt im Krankenhaus. Vergleich der Erfahrungen und Wahrnehmungen deutscher und türkischsprachiger Patientinnen sowie des Klinikpersonals zur Versorgungssituation in der Gynäkologie [Patient orientation in the context of socio-cultural diversity in the hospital. A comparison of experiences and perceptions of German and Turkish speaking patients as well as the hospital personnel about the healthcare situation in gynaecology]', PhD thesis, Technische Universität Berlin, Fakultät VIII – Wirtschaft und Management, http://edocs.tu-berlin.de/diss/2002/borde_theda.htm

Borde, T., David, M. and Kentenich, H. (2002) 'Erwartungen und Zufriedenheit deutscher und türkischsprachiger Patientinnen im Krankenhaus: eine vergleichende Befragung in einer Berliner Frauenklinik [What Turkish-speaking women expect in a German hospital and how satisfied they are with healthcare during their stay in a gynaecological hospital in Berlin – a comparative approach]', *Gesundheitswesen*, vol 64, nos 8–9, pp476–485

Borde, T., Braun, T. and David, M. (2003) *Unterschiede in der Inanspruchnahme klinischer Notfallambulanzen durch deutsche Patienten/innen und Migranten/innen – Problembeschreibung, Ursachenanalyse, Lösungsansätze [Differences in the Utilization of Clinical Emergency Departments in German and Migrant Patients – Problem Definition, Cause Analysis, Approaches]*, Final report of the research project funded by the Federal Ministry for Education and Research and the Umbrella Organization of the Statutory Health (No: 01 GL 0009), Berlin

Burchard, E. G., Ziv, E., Coyle, N., Gomez, S. L., Tang, H., Karter, A. J., Mountain, J. L., Perez-Stable, E. J., Sheppard, D. and Risch, N. (2003) 'The importance of race and ethnic background in biomedical research and clinical practice', *New England Journal of Medicine*, vol 348, no 12, pp1170–1175

Collins, G. S. and Johnson, M. R. D. (2007) 'Addressing ethnic diversity in health outcome measurement: A systematic and critical review of the literature', www2.warwick.ac.uk/fac/med/research/csri/ethnicityhealth/research/ethtom_summary/, accessed 29 September 2008

Culley, L., Hudson, N., Rapport, F., Katbamna, S. and Johnson, M. R. D. (2006) 'British South Asian communities and infertility services', *Human Fertility*, vol 9, no 1, pp37–45

Culley, L., Hudson, N. and Rapport, F. (2007) 'Using focus groups with minority ethnic communities: Researching infertility in British South Asian communities', *Qualitative Health Research*, vol 17, no1, pp102–112

David, M. and Borde, T. (2001) *Kranksein in der Fremde? Türkische Migrantinnen im Krankenhaus [Being Ill Abroad? Turkish Migrant Women in Hospital]*, Mabuse-Verlag, Frankfurt am Main

David, M., Borde, T. and Kentenich, H. (2000) 'Knowledge among German and Turkish women about specifically female bodily functions, contraception, preventive medical examinations and menopause', *Ethnicity and Health*, vol 5, pp101–112

David, M., Schwartau, I., Pant, H. A. and Borde, T. (2006) 'Emergency outpatient services in the city of Berlin: Factors for appropriate use and predictors for hospital admission', *European Journal of Emergency Medicine*, vol 13, pp352–357

Department of Health (2005) *A Practical Guide to Ethnic Monitoring in the NHS and Social Care*, DH / Health and Social Care Information Centre / NHS Employers, London

Drew, D. (1980) 'The politics of statistics', in Runnymede Trust (ed) *Britains' Black Population*, Heinemann, London

Erens, P., Primatesta, P. and Prior, G. (2001) *Health Survey for England: The Health of*

Minority Ethnic Groups, Department of Health, London

European Union (1997) *The Amsterdam Treaty* http://europa.eu/scadplus/leg/en/lvb/ a10000.htm, accessed 1 October 2008

Federal Statistical Office (2006) 'Bevölkerung mit Migrationshintergrund – Ergebnisse des Mikrozensus 2006 [Population with migrant background – Results of the micro-census 2006]' Fachserie 1 Reihe 2.2 – 2006, https://www-ec.destatis.de/, accessed 28 November 2008

Fitzpatrick, J., Jacobson, B. and Aspinall, P. (2005) *Indications of Public Health in the English Regions*: *Vol 4, Ethnicity and Health*, Association of Public Health Observatories, London

Gacinski, L., Yüksel, E. and Kentenich, H. (2002) 'The unfulfilled desire for a child of oriental couples: Infertility counseling and treatment of Turkish immigrants in Germany', in Strauss, B. (ed) *Involuntary Childlessness*, Hogrefe & Huber Publishers, Seattle, WA

Hippisley-Cox, J., Coupland, C., Vinogradova, Y., Robson, J., Minhas, R., Sheikh, A. and Brindle, P. (2008) 'Predicting cardiovascular risk in England and Wales: Prospective derivation and validation of QRISK2', *British Medical Journal*, vol 336, pp1475–1482

Hunt, S. and Bhopal, R. (2003) 'Self-reports in research with non-English speakers', *British Medical Journal*, vol 327, pp352–353

Johnson, M. R. D. (1998) 'Ethnic monitoring: Bureaucratic construction of a minority entity or identity', in C. Williams, H. Soydan and M. Johnson (eds) *Social Work and Minorities*, Routledge, London

Johnson, M. R. D. (2001) 'Ethnic monitoring and nursing', in L. Culley and S. Dyson (eds) *Ethnicity and Nursing Practice*, Palgrave Macmillan, London

Johnson, M. R. D. (2006) 'Engaging communities and users: Health and social care research with ethnic minority communities', in J. Y. Nazroo (ed) *Health and Social Research in Multiethnic Societies*, Routledge, London

Johnson, M. R. D., Wright, A., Jeffcoat, M.-A. and Petherick, R. (1996) 'Local authority occupational therapy services and ethnic minority clients', *British Journal of Occupational Therapy*, vol 59, no 3, pp109–114

Kentenich, H. (2002) 'Patients in migration', in J. Boivin and H. Kentenich (eds) *Guidelines for Counselling in Infertility*, European Society of Human Reproduction and Embryology (ESHRE) Monographs Oxford University Press, Oxford

Lin, S. S. and Kelsey, J. L. (2000) 'Use of race and ethnicity in epidemiologic research: Concepts, methodological issues and suggestions for research', *Epidemiologic Reviews*, vol 22, no 2, pp187–202

Lloyd, C. E., Mughal, S., Sturt, J., O'Hare, P. and Barnett, A. H. (2006) 'Using self-complete questionnaires in a South Asian population with diabetes: Problems and solutions', *Diversity in Health & Social Care*, vol 3, no 4, pp245–251

Lloyd, C. E., Johnson, M. R. D., Mughal, S., Sturt, J. A., Collins, G. S., Roy, T., Bibi, R. and Barnett, A. H. (2008) 'Securing recruitment and obtaining informed consent in minority ethnic groups in the UK', *BMC Health Services Research*, vol 8, no 68, pp1–9

Macpherson, W. (1999) 'The Stephen Lawrence Inquiry: Report of an Inquiry', Home Office, London, www.archive.official-documents.co.uk/document/cm42/4262/sli-00.htm, accessed 29 September 2008

Maschewsky-Schneider, U. and Fuchs, J. (2000) 'Brauchen wir in der Forschung zu Migration und Gesundheit besondere methodische Zugangsweisen? Was wir aus der Frauenforschung lernen können. [Do we need specific methodological approaches in researching migration and health? What we can learn from gender research]', in M. David, T. Borde and H. Kentenich (eds) *Migration – Frauen – Gesundheit. Perspektiven im europäischen Kontext [Migration – Women – Health. Perspectives in the European Context]*, Mabuse-Verlag, Frankfurt am Main

McLean, C. A. and Campbell, C. M. (2003) 'Locating research informants in a multi-ethnic community: Ethnic identities, social networks and recruitment methods', *Ethnicity & Health*, vol 8, pp41–61

O'Hare, P., Raymond, N. T., Mughal, S., Dodd, L., Hanif, W., Ahmed, Y., Mishra, K., Jones, A., Kumar, S., Szczepura, A., Hillhouse, E. W. and Barnett, A. H. (2004) 'Evaluation of enhanced diabetes care to patients of South Asian ethnicity: The United Kingdom Asian Diabetes Study (UKADS)', *Diabetic Medicine*, vol 21, pp1357–1365

Pette, G. M., Borde, T. and David, M. (2004) 'Kenntnis über die Diagnose und Therapie ihrer Erkrankung bei deutschen und türkischstämmigen Patientinnen vor und nach einem Krankenhausaufenthalt [Turkish and German patients' current knowledge of diagnosis and therapy at the beginning and at the end of the inpatient stay]', *Turkish German Gynecol Assoc*, online version, vol 5, no 2, pp123–130

Public Health Sector Group (2000) *Ethnicity Profiling in Primary Care: The Princes Park Health Centre Model*, Public Health Sector Group, School of Health and Human Sciences, Liverpool John Moores University, Liverpool

Royal Statistical Society (1983) 'Sources of statistics on ethnic minorities', *Journal of the Royal Statistical Society*, vol 146, pp99–114

Tutton, R. (2007) 'Opening the white box: Exploring the study of whiteness in contemporary genetics research', *Ethnic and Racial Studies*, vol 30, no 4, pp557–569

Winslow, W. W., Honein, G. and Elzubeir, M. A. (2002) 'Seeking Emirati women's voices: The use of focus groups with an Arab population', *Qualitative Health Research*, vol 12, pp566–575

Yüksel, E., Siemann, A. and Kentenich, H. (1996) 'Kinderlosigkeit Turkischer Paare in der Migration. Ein multifaktorielles Problem [Childlessness among Turkish migrant couples. A multi-factorial problem]', *Geburtshilfe und Frauenheilkunde*, vol 56, pp188–190

5

What Difference Does Our Difference Make in Researching Infertility?

Yasmin Gunaratnam

A young Black Caribbean mother reflecting upon her involvement in a research study tells the Black Caribbean interviewer 'Some of the questions you could actually feel was a white person asking them, and some of them were just so stupid that you could get the feeling that somebody was trying to get inside black people to find out what it is like' (Phoenix, 2001, p207). In Bradford, a British-born Indian Sikh researcher seeking access to Pakistani youth is met with suspicion from community leaders and is routinely met with the questions 'You don't sound like you are from around here? Who do you work for? Who's funding the project, the government?' (Sanghera and Thapar-Bjorkert, 2008, p554). An Indo-Canadian researcher investigating reproductive decision making among women in India finds her training and supervision within a British university woefully inadequate. As a 'diasporic', young and junior researcher, her authority and identity were questioned and often undermined in the field. She concludes that her white supervisors' 'fieldwork warnings and advice [were] being tailored to assumptions about western privilege and power' (Henry, 2007, p73).

These brief examples provide some insight into how our differences can matter in research. They also show some of the assumptions and difficulties that can confront researchers. For instance, although we may recognize that categories such as ethnicity, race and culture are socially constructed and relational (see Chapter 2), the notion of absolute and reified difference conveyed in the first example shows how in everyday life 'essentialist identities continue to be invoked and [are] often deeply felt' (Calhoun, 1994, p14, quoted in Ang, 2000, p2). In the remaining examples, we are brought face-to-face with the 'power-geometries' (Massey, 1999) of different research spaces, in the 'field' and at 'home'.

In Bradford, it was the historical context of this English, northern, multi-cultural city at the time of the research that is significant. According to the researchers, it was the 'climate of fear and suspicion' (Sanghera and Thapar-Bjorkert, 2008, p552) in Bradford following urban unrest and concerns about Islamic fundamentalism, rather than ethnicity per se, that was felt to impact upon the researcher's access to potential research participants. Marsha Henry's (2007) student experiences in the final example speak to the interplaying between social identities and institutional relationships, as well as highlighting the complexities of fieldwork for those with hyphenated ethnic identities, and in different national contexts. Here, significant issues relate to the quality and the nature of support and supervision that are provided to fieldworkers, and to the care that research teams take in addressing the implications of their varying differences and institutional locations throughout the research process (Dyck et al, 1995). The difference that our difference can make in researching a sensitive topic such as infertility is multi-layered and multi-sited.

The epistemological frameworks, methodological form and the aims of our research are of course also significant in framing how we take account of our difference. There is a surprisingly long tradition in survey research, for example, where the technical management of the social characteristics of the interviewer has been a part of efforts to standardize interviews and to eliminate 'interviewer related error'. This tradition has included efforts to experiment with and manage what have been called 'race-of-interviewer-effects', where the race/ethnicity of the interviewer has been seen as affecting the reliability of opinion-related data (see Schuman and Converse, 1971; Schaeffer, 1980; Schaeffer and Maynard, 2002). In contrast, and largely in qualitative research, feminist scholarship has pioneered and transformed how difference has been approached methodologically, ethically and epistemologically, with attention being given to research as a psychosocial interaction, underwritten by historical and inter-subjective relations of power that need to be addressed and accounted for (Harding, 1986; Haraway, 1988; Phoenix, 2001). 'Feminist research' writes Edwards 'should not just be *on* women but *for* women' (1990, p479, emphasis in original).

Despite the differences between quantitative and qualitative research, two inter-related themes that are central to methodological discussions of how the social identities of researchers matter are those of commonality and difference or being an 'insider' or 'outsider'. Much of the early methodological writing on race and ethnicity suggested that shared race/ethnicity was a resource in research and that difference could impair rapport and communication. Summarizing the dominant themes in the literature, Rhodes has observed how:

Closeness of identity and, in particular shared racial identity is generally presumed to promote effective communication between researcher and subject and, conversely, disparate identity to inhibit it. (1994, p550)

Ongoing methodological research and debates among feminist scholars and researchers from racialized minorities have challenged such understanding of difference in research, whilst simultaneously undermining 'the notion of truth and objectivity and ... the notion of the value-free, invisible researcher and the passive, accessible research subject' (Alexander, 2006, p399). The scholarship that has emerged has drawn attention to issues of ethnic and cultural mixity

(Song and Parker, 1995; Henry, 2007), and the interplaying of ethnicity with other social identities (Bhavnani, 1993; Phoenix, 2001), in time and place (Gunaratnam, 2003; Sanghera and Thapar-Bjorkert, 2008).

Recognizing the inevitable complexity of difference in research, Nast has used the metaphor of 'betweenness' to capture the dynamism and contingency of difference. She writes:

> *Even where differences in a field are small, because we are positioned simultaneously in a number of fields we are always, at some level, some-where, in a state of betweenness, negotiating various degrees and kinds of difference – be they based on gender, age, class, ethnicity, 'race', sexuality, and so on. Betweenness thus implies we are never 'outsiders' or 'insiders' in any absolute sense.* (1994, p57, quoted in Dyck et al, 1995, p616)

In this chapter, I want to retain the ethos of Nast's focus on 'betweenness'. I find the dualism inherent in the language and spatial metaphors of 'insiders' and 'outsiders' limiting, not least because it instigates an absolute and somewhat harsh categorization and splitting of people and their relationships (not dissimilar to processes of racism themselves). This splitting is especially constraining when it comes to recognizing unpredictable connections and a depth of dialogue and attentiveness across difference (for examples see Liebow, 1967; Lather and Smithies, 1997; Inhorn, 2007).

In what follows I will keep the idea of betweenness in mind as I first address the epistemological and ethical challenges of accounting for difference, and then give attention to experiences in three main domains of research: the social and historical context; accessing research participants; and in fieldwork. This discussion will draw upon a range of literature and areas of research, largely because methodological writing on how difference matters in infertility research is, at present, scant.

Politics and poetics

Broadly speaking, both infertility and racialized difference are threatening or 'sensitive' topics (see Chapter 4), raising dual concerns about emotions and politics, and ultimately raising challenging questions about the nature of identity itself. While recognizing that infertility is an experience in which identity can become both more salient, and more vulnerable and contested (Letherby, 2002), I am also acutely aware that the imagination of race, ethnicity and culture as 'things' that we can know and can identify a priori, and that can separate out 'Asian' from 'African' or 'Polish' experiences of infertility, are ways of thinking that while habitual and meaningful in real life, *always* fail to do justice to the sheer diversity, inventiveness and biographical idiosyncrasy of what it means to be human and vulnerable.

I am reminded here of Khaled Hosseini's (2003) beautiful and poignant novel *The Kite Runner*, set in Afghanistan, the US and Pakistan. The novel, layered by culture, religion and caste, portrays the turbulent moral struggle of Amir to find a 'way to be good again' following his betrayal of his childhood friend and servant, Hussan. In the US, Amir and his wife Soraya are diagnosed with 'unexplained infertility' and their subsequent IVF (in vitro fertilization) treatment is unsuccessful. In charting their movement through the medical system, Hosseini evokes the slow losses and increasing emotional weight of infertility in the couple's lives. At night, with Soraya sleeping beside him, we find Amir feeling the emptiness of Soraya's womb 'like it was a living breathing thing. It had seeped into our marriage … into our laughs, and into our lovemaking… I'd feel it rising from Soraya and settling between us… Like a newborn child' (2003, p165).

Within the narrative, Hosseini embeds the experience of infertility in the context of the lives of an Afghani migrant community, where the fatalism of 'God knows best' from Soraya's mother and her father's aversion to adoption can perhaps be grasped as being inscribed by both generation and culture. Yet Hosseini never allows us the emotional distancing or laziness of the 'it's cultural' interpretation. By enfolding infertility into the biographical density of guilt, disappointment and loss in Amir's life we are also brought close to naked human experiences of vulnerability and suffering and the intensely personal emotional meanings of infertility for Amir. In Hosseini's hands there is a blurring of sharp edges and we are never entirely sure of what is cultural, what is universal, and what is Amir.

Literature is perhaps more efficacious than empirical research in conveying the mysteries and lived acrobatics of difference (Gunaratnam, 2007), yet I am also suggesting that we need to be as caring, forensic and delicate as Hosseini is in the ways that we think about, investigate and account for difference in our own research. Would our research stand up to the 'Amir test'?: if Amir were involved in one of our research projects, would our methodological tools and practices be attuned to any of the nuances of his experiences of infertility? By posing such a challenge, I do not mean to suggest that we should give up on the empirical quest for describing and understanding manifestations of difference – because it is ultimately always moving on and is intangible. As Les Back comments: 'The task, it seems to me, is to pay truth the courtesy of serious effort without reducing the enigmatic and shifting nature of social existence to caricature and stereotype' (2007, p153).

In what follows, I am therefore very much concerned with the ethics and the artistry of our methodological practices; the ways in which our actions and interpretations of difference can both leave open 'a possibility for humanity' (Duneier and Back, 2006, p553) and 'allow us to become answerable for what we learn how to see' (Haraway, 1988, p583).

Ethnicity and the social and cultural context of research

There is no doubt that there is an increased interest in ethnicity and cross-cultural study in contemporary social research, in part due to a significant expansion of the disciplines concerned with race, ethnicity and culture, and in part due to the emergence of new generations of scholars from racialized minorities (Alexander, 2006). This interest is also inseparable from the wider social and cultural context of research and from trenchant critiques of ethno-centrism, neo-colonialism and racism in research (Clifford and Marcus, 1986; Stanfield and Dennis, 1993; Visweswaran, 1994; Tuhiwai Smith, 1999; Twine and Warren, 2000).

In reviewing the anthropological 'virgin birth' debates (about Australian Aboriginals and Trobriand Islanders who were said to be unaware of physiolog-ical paternity), Franklin (1997) has shown how accusations of racism and a Judaeo-Christian cultural logic were core features of the debates that failed to problematize beliefs about the 'facts of life' in Western cultures. At the level of global interrelations, Feldman-Savelsberg (2002) has identified 'demographic prejudice' as preventing researchers and policy makers from even recognizing infertility in regions such as the Cameroon that are perceived to be 'overpopu-lated'. And at the level of research findings, researchers have critiqued what have been seen as simplistic (Lawrence, 1982) and 'boring travesties' (Back, 2007, p154) of portrayals of the cultural life and practices of those from different ethnic and cultural backgrounds.

This wider context of research has been linked to difficulties in gaining access to, and the cooperation of, socially marginalized communities and research participants, where there can be resentment or unease about research that is overtly concerned with ethnicity and difference and where the topic under inves-tigation is 'sensitive' and subject to stigmatization (Elam and Chinouya, 2000). Some 'minority ethnic' communities, we are told, can feel over-researched (Darlow et al, 2005); jaded by the lack of tangible benefits from their involve-ment in research (Bengston et al, 1977; Butt and O'Neil, 2004); and can be suspicious of the motives of researchers who are perceived to 'parachute' into communities that they have instrumental and short-term connections with (Sanghera and Thapar-Bjorkert, 2008). An implicit assumption is that differ-ences of ethnicity in research can feed into such anxieties, sensitivities and suspicion, increasing the difficulties that researchers face in getting to research participants.

The timing of research can play a critical part in such dynamics. The American anthropologist Marcia Inhorn (2004) found her research proposal on male infertility in Egypt rejected repeatedly by male Egyptian security officers who presided over her scholarship exchange programme. Attributing this rejec-tion to a gendered 'ego security' about the topic of male infertility, Inhorn further suggests that:

> *Clearly, the rejection of my proposed study ... had multiple sources, but could largely be summed up in one word: politics. These politics certainly included the tense US–Egyptian political relations, particularly following September 11th; the trend in Egypt toward heightened scrutiny of all Western researchers and non-governmental organizations ...* (p2103)

Although Inhorn's experience may be an extreme one, it is clear that not enough consideration is given by researchers or funding bodies to 'politics' in its varying forms, and to the increased demands upon methodology, time and financial resources it can take to build up trust with communities and gatekeepers in order to access and recruit socially marginalized research participants (Culley et al, 2007). In her ethnographic research on IVF in hospitals in the Middle East, where 'infertility and IVF are shrouded in layers of secrecy and social suffering' (Inhorn, 2004, p2097), Inhorn found that she had to adapt her methodological practices, abandoning tape-recorded interviews and providing detailed oral and written reassurances of confidentiality and anonymity to engender more trusting relationships and to facilitate the participation of local women in the research.

In terms of ethics and politics, concerns that research on race and ethnicity, like feminist research, should be 'for' rather than simply 'on' different ethnic and cultural groups has led to attention to reciprocity, activism and to how research might be of practical benefit to local communities. Initiatives have included project steering groups with local community representatives and/or the 'users' of research; accredited research training for community researchers (Kai and Hedges, 1999); the production of information for communities on infertility in different languages and media (Culley et al, 2007); and involving research participants in university teaching (Duneier, 1999).

There are obvious dangers in romanticizing such initiatives and perspectives – the social exchange system is highly unstable and ambiguous (Leifer, 1988) and there can only ever be intuitive impressions of both what is exchanged and its value (Gunaratnam, 2003). Nevertheless, to face, tussle with, or try to do something that makes a difference to social marginalization and racism is also an integral part of how our difference is implicated and can have consequences in research. As Elam and Chinouya have also pointed out:

> *When people are not treated as citizens in many aspects of their lives it is unlikely that they will perform as citizens in other respects. People have to believe that research will benefit their community. If the community is poorly treated or does not receive feedback or see any benefits, this belief will be difficult to sustain.* (2000, p46)

Ethnicity and access

In their study of infertility among South Asian groups in three English cities, Culley et al (2007) were mindful that the social identities of the lead researchers as white women could have a detrimental effect upon their access to research participants. Yet as the research progressed they found that:

> *... the status of the team members as 'experts', rather than their ethnicity, became the social identity of most significance to those individuals who were important 'gatekeepers'. In addition, an association with the locally known university and funding from a major public body seemed to assure the gatekeepers of the trustworthiness of the research team. These gatekeepers were vital in the acceptance of the researchers and in gaining access to*

groups by whom the core researchers might have been perceived negatively as 'outsiders'. (p107)

That racial and ethnic commonalities are not determining factors in accessing research participants has been supported by the accounts of fieldworkers from racialized minorities (Thapar-Bjorkert and Henry, 2004; Henry, 2007; Sanghera and Thapar-Bjorkert, 2008). What is interesting about this literature is that it points to the shifting salience of constructions of race/ethnicity as a part of a complicated and ambivalent positionality that is connected to institutional status, timing, and the micro-politics of gatekeeping and patronage within particular localities.

In a study on reproductive decision making in India (Thapar-Bjorkert and Henry, 2004; Henry, 2007), Henry experienced a very different set of research relations in accessing research participants to those described by Culley et al (2007). For Henry, the duality of her ethnic identity, her youth, and position as a junior researcher, mediated privileges of occupational status and expertise so that in many settings she felt that she was seen as a 'daughter-cum-apprentice' (2007, p74), a position of occupational and relational immaturity that she describes as 'straddling' and moving between insider and outsider categorizations.

In the Bradford study discussed earlier (Sanghera and Thapar-Bjorkert, 2008), the fieldworker Gurchathen Sanghera's account of his experiences with gatekeepers charts similar complex positionings. With some gatekeepers, Sanghera felt that his Asianness was enough to make him 'apna' (one of us), ensuring access, but it also led to problematic assumptions about commonalities of experiences of racism. For other gatekeepers, Sanghera's status as a researcher, raised outside of Bradford and identified with a predominantly white and middle class profession and institution, in combination with local feelings of being 'researched out', positioned him as not being a part of an imagined community of 'us'.

Sanghera describes an incident when he had telephoned a potential gate-keeper at home. On passing the phone to her husband, he heard the gatekeeper's wife say in Punjabi, 'Koi-y ghora wha' (it's some white man). It is clear from Sanghera's account that he felt that this comment served to distance him from the participant, but this was not a permanent or entirely racialized estrangement. Here is the description of the phone call and subsequent developments in Sanghera's relationship with the gatekeeper:

With that remark, the researcher was positioned as an outsider – a white male – and the potential gatekeeper seemed quite guarded when he spoke to the researcher; he asked, 'Do you work for the media? How did you get my home number?' When the researcher and gatekeeper met in person, he was very hospitable, less guarded and generally helpful. This may have been because … there were obvious commonalities – indeed he commented that 'It's the first time I've come across an Asian researcher, it's quite rare… You've lived in Bradford? That's good, and then you understand the complexities of it all.' We would contend that the gatekeeper re-positioned the researcher from an outsider to more of an insider. (p555)

The varied and evolving influences of ethnicity on relationships of access have significant practical and ethical implications, especially for multi-ethnic teams where research assistants and community or bilingual researchers are so often on the 'front line' of research interactions (Dyck et al, 1995). As more of us write about our research experiences, it is becoming increasingly apparent that the institutional organization of research can result in differential and racialized vulnerabilities and also 'emotional labour' (James, 1989) in research. What I mean by this is that the personal identity and motivations of minority or community researchers can be particularly exposed and up for questioning, whilst those in positions of greater authority and occupational security can be 'invisible researchers' (Dyck et al, 1995), shielded or distanced from the personalized intrusions and emotionally charged negotiations of access.

It is also the case that, as Nayak has observed, 'while some principle investigators have willingly deployed minority researchers for data collection and interpretation, this has not always resulted in write-up, publication and a change in status for these workers' (2006, p413). An all too familiar and politically charged irony here is that a commitment to equality and the involvement of marginalized groups in research can be compromised by a lack of attention to difference and its effects within the organization and hierarchies of research itself.

Doing research: To match or not to match?

In research that involves ethnic and cultural difference, researchers will often consider the advantages and disadvantages of shared ethnicity between the researcher and the research participant(s), particularly when language is also an issue. There is no robust empirical evidence to suggest that shared race or ethnicity yields better/richer (more 'valid' or 'reliable') data. Nevertheless, it is important to recognize that a lack of attention to social, cultural and linguistic differences in research can lead to significant misunderstanding (Riessman, 1987). There is also a growing body of survey research that points to the 'response effects' of racial and ethnic difference between interviewers and research participants, where questionnaire responses to opinion questions have been found to vary with controlled changes to the race/ethnicity of interviewers and research participants (see Gunaratnam, 2003, ch 3). These data are far from conclusive however (see Ipsos Mori, 2002), particularly when examined in relation to the aims of research, its institutional organization and timing (Gunaratnam, 2003). This inconclusiveness is indicative of the inherent difficulties of identifying and measuring what are essentially situational and inter-subjective relations between researchers and research participants, involving conscious and unconscious emotions, and this is a point I will return to later.

Although research interactions are highly complicated and there is no firm empirical basis to support the benefits of shared ethnicity in research, ethnic matching has been promoted as a practice of 'cultural competence' in research. For instance, in research involving interviews it has been argued that matching:

encourages a more equal context for interviewing which allows more sensitive and accurate information to be collected. A researcher with the same ethnic background as the participant will possess ... an insider/emic view ... will have more favourable access conditions and the cooperation of a large number of people ... and a genuine interest in the health and welfare of their community. (Papadopoulos and Lees, 2002, p261)

Such a rationale for the benefits of using ethnic matching is both problematic and flawed. It assumes that ethnicity will have primacy in research interactions and that it can be used, indeed exploited, as 'methodological capital' (Gallagher, 2000) to gain more 'accurate' data and better access to and cooperation from research participants. Most significantly, it is based upon a logic of commonality as emotional and ethical unity, where race, ethnicity and/or culture are imagined as imbuing research interactions with levels of communication, trust and care that precede the research relationships themselves.

Henry's experience of using a local gatekeeper as her interpreter in her research in India shows how relations of commonality and difference can also be mediated by differences of language and research roles. For Henry, her position as Indo Canadian and a stranger to the community she was studying is acknowledged as being a possible advantage in fieldwork, enabling women to disclose sensitive information. But Henry also suggests that while the local connections of her interpreter did not inhibit women disclosing information about their fertility practices, she suspected that some research participants projected idealized versions of their marriage in interviews, fearful that accounts of marital discord may have been recycled as 'neighbourhood gossip' (2007, p370).

Psychosocial approaches have been innovative and productive in developing methods and forms of interpretation and analysis that recognize how we might use the intuitive impressions and experience of researchers as an 'instrument' of knowing (Hollway, 2007), to examine the effects of social and emotional differences in research. Research that I am currently involved with – that is examining becoming a mother for the first time among white British, Bangladeshi and African-Caribbean women in Tower Hamlets in London – provides an interesting case study example of methodological attempts to address matters of difference in research from a psychosocial perspective. The study has used a combination of repeat narrative interviews (supplemented by detailed field notes describing interview dynamics), with psychoanalytic observation of mothers and their babies, using trained psychoanalytic observers, with observations taking place once a week over a 12-month period.

Analysis of both the interview and observational data has involved different forms of group analysis, where attention has been given to the effects and interpretations of the social difference between the researchers and observers (i.e. mainly with regard to ethnicity, culture and class) and the women and babies. This group work has enabled a more critical and systematic interrogation of difference, addressing concerns about the limits of reflexivity in research where it has been argued that individual researchers are only ever able to 'explicate the processes and positions we are aware of being caught up in' (Reay, 1996, p443).

Cathy Urwin (2007), a child psychotherapist who led the weekly observation seminar group to process the observational data for the study has described how group analysis was used to generate an understanding of emotional interactions and a critical examination of difference:

> *Once emotional impact had been registered, we then began to elucidate the source of the emotional experience in the observational material, to encourage the group to objectify intuition with evidence, moving gradually into the examination of sequences of behaviour. In this process we paid particular attention to the observer's and the group's reactions to observations of unfamiliar practices, and to the impact of cultural and ethnic differences …*

> *… detailed notes were kept of the discussion of each observation. These played an important part in the triangulation or validation required in the research process. They provided a record of the group's attempts to find meanings in the observation that go beyond the viewpoint of the singular observer, noting alternative possibilities. We referred back to them when we next heard about the particular mother, and modified inaccuracies, elaborating when necessary. This record facilitated our being able to note and keep track of recurrent themes.* (pp245–246)

Analysis of the findings of the research is ongoing at the time of writing. However emerging themes suggest that the significance of ethnic and cultural differences between the researchers/observers and the mothers was mediated by class and age; the development of the research relationships over time; by the physical spaces of where the interviews/observations took place and who else was in them; and by unconscious identifications and defences.

The complex realities of doing research are therefore somewhat at odds with the assumptions made in ethnic matching approaches. Such assumptions also do a great disservice to research that makes connections across difference, is sensitive to the specificities and patterns of locality, and yet leaves open 'a possibility for humanity' (Duneier and Back, 2006, p553). Inhorn's (2007) ethnographic research of men's experiences of masturbation and semen collection in IVF clinics in Lebanon is a powerful example of what research can achieve when there are differences of gender, ethnicity, culture and religion between researchers and participants. Through observation and interviews, Inhorn shows how the universally fraught process of semen collection, with its demands of sexual performance, is further characterized by the interplaying of the architecture of IVF clinics, variations in cultural sensitivities towards masturbation among institutions and professionals, and the 'complex meanings of sin, guilt and even illicit pleasure for IVF-seeking men in the Muslim world' (p39).

Although Inhorn does not discuss how her difference affected her research, we are left with the possibility that she was able to generate such rich narrative accounts from the men, not despite, but rather *because* of her difference. Such an interpretation takes account of the morality involved in exposing difference and vulnerability to an Other. Let me explain. There is a growing body of research that shows how experiences of infertility can be inscribed by cultural and religious mores (see Sewpaul, 1999), but it is much more difficult to get a sense of how individuals and couples are more than passive victims or puppets of cultural and religious attitudes and beliefs. In the following extract from an interview between Inhorn and a Lebanese Shia Muslim construction worker, we get a rare glimpse of how a self-conscious gendered vulnerability can involve a critical orientation towards the perceived gendered restrictions of religion and culture. What is striking about this narrative is its intense moral critique of what are felt to be 'Arabic' and Islamic attitudes towards male sexuality, which in turn are connected to infertility:

> *Arabs don't have a reasonable attitude toward sex. The problem is, the mothers are always telling their children, especially in the Muslim community, 'This is no good.* Haida haram! *[i.e. this is sinful].' Just to think about sexual matters is wrong. Ever since I was young, my mother used to 'shush' me if I even brought it up. The way I was raised and the things I was taught may have affected my fertility now. I had no education on sexuality. Everything was 'no good'. It was a big mistake that I wasn't taught. So, in cases where I would have an erection as a teenager, I wouldn't know what to do, because I wasn't taught... All over the world, every teenager goes through this experience, and at this age, they start masturbating. I'm asking myself, maybe due to excess masturbation, maybe this affected my sexual life and my fertility later on.* (2007, p44)

Without an account of Inhorn's interpretations of how difference affected her research relationships, it is difficult to provide a detailed exploration of the social and inter-subjective dynamics that might have produced such a narrative. Nevertheless, what I want to draw attention to is how the account is relevant to the concept of 'dialogism', pioneered by the literary theorist Mikhail Bakhtin. For Bakhtin (1986), rather than difference being a barrier to communication, it can be the very site of dialogue and understanding: it can provoke radical interrogation and reflection on what is taken for granted and it can enable tentative or marginalized perspectives to be spoken and tried out. For Bakhtin 'A meaning only reveals its depths once it has encountered and come into contact with another, foreign meaning: they engage in a kind of dialogue, which surmounts the closedness and one-sidedness of these particular meanings, these cultures' (1986, p7). Most importantly, Bakhtinian dialogue assumes an ethical relationship in difference that raises fundamental questions about our relationships and responsibilities to others.

In his work on illness and care, the sociologist Frank (2004) outlines questions raised by dialogism that are also highly relevant to difference and infertility researchers. He asks:

> *Do I see the tensions that the other's vulnerabilities create? Do I recognize what the other is having to hold together, to carry on at all, and his or her fear of life coming apart? Then what part does this other cast me in, in this drama? What does he or she fear I will play, and what part does he or she hope I might play? What is my obligation ... to be who this ... vulnerable person needs me to be?* (p141)

Conclusion

In her discussion of translation and interpretation among non-English speaking Hispanic women in the US, Noreen Esposito (2001) opens her account with recourse to a Simple Life fantasy. A fantasy that I am sure is not uncommon – in its different versions – to many social researchers. For Esposito it is this: 'How much easier it would be in cross-language research if we could understand the participants' own words and nuances without the need for third party translation' (p568). For those concerned with questions of ethnicity and difference in research, the field can appear daunting – methodologically complicated, resource intensive, and ethically fraught: how much easier it would be if our difference didn't have an impact upon how we do research and the findings we generate?

Of course, the irony of such a fantasy is that *all* social research involves ethnicity and difference. It is just that ethnicity can become more noticed and provocative where there are racialized differences between researchers and research participants and where the research topic is felt to be 'sensitive'. Yet, as I have discussed, difference matters deeply in research and not just as a methodological problem. Attentiveness to difference can enrich our conceptual frameworks, lead to innovations in methodology and can push and pull us to face our ethical responsibilities. Where would we be without it?

References

Alexander, C. (2006) 'Introduction: Mapping the issues', *Ethnic and Racial Studies*, vol 29, no 3, pp397–410

Ang, I. (2000) 'Identity Blues', in P. Gilroy, L. Grossberg and A. McRobbie (eds) *Without Guarantees: In Honour of Stuart Hall*, Verso, London

Back, L. (2007) *The Art of Listening*, Berg, Oxford

Bakhtin, M. (1986) 'Speech genres and other late essays', in C. Emerson and A. Holquist (eds) *The Dialogic Imagination: Four Essays by M.M. Bakhtin*, University of Texas Press, Austin, TX

Bengston, V., Grigsby, E., Corry, E. and Hurby, M. (1977) 'Relating academic research to community concerns: A case study in collaborative effort', *Journal of Social Issues*, vol 33, no 4, pp74–92

Bhavnani, K.-K. (1993) 'Tracing the contours of feminist research and feminist objectivity', *Women's Studies International Forum*, vol 6, no 2, pp95–104

Butt, J. and O'Neil, A. (2004) *Let's Move On: Black and Ethnic Minority Older People's Views on Research Findings*, Joseph Rowntree Foundation, York

Calhoun, C. (1994) 'Social theory and politics of identity', in C. Calhoun (ed) *Social Theory and Politics of Identity*, Blackwell, Oxford

Clifford, J. and Marcus, G. (eds) (1986) *Writing Culture: The Politics and Poetics of Ethnography*, University of California Press, Berkeley, CA

Culley, L., Hudson, N. and Rapport, F. (2007) 'Using focus groups with minority ethnic communities: Researching infertility in British South Asian communities', *Qualitative Health Research*, vol 17, no 1, pp102–112

Darlow, A., Bickerstaffe, T., Burden, T., Green, J., Jassi, S., Johnson, S., Kelsey, S., Purcell,

M., South, J. and Walton, F. (2005) 'Researching Bradford: A review of social research on Bradford District', Joseph Rowntree Foundation / Leeds Metropolitan University, www.jrf.org.uk/bookshop/eBooks/1859354262.pdf, accessed 12 December 2006

Duneier, M. (1999) *Sidewalk*, Farrar, Strauss and Giroux, New York

Duneier, M. and Back, L. (2006) 'Voices from the sidewalk: Ethnography and writing race', *Ethnic and Racial Studies*, vol 29, no 3, pp543–565

Dyck, I., Lynam, J. and Anderson, J. (1995) 'Women talking: Creating knowledge through difference in cross-cultural research', *Women's Studies International Forum*, vol 18, nos 5/6, pp611–626

Edwards, R. (1990) 'Connecting method and epistemology: A white woman interviewing black women', *Women's Studies International Forum*, vol 13, no 5, pp477–490

Elam, G. and Chinouya, M. (2000) *Feasibility Study for Health Surveys among Black African Populations Living in the UK: Stage 2: Diversity among Black African Communities*, National Centre for Social Research, London

Esposito, N. (2001) 'From meaning to meaning: The influence of translation techniques on non-English focus group research', *Qualitative Health Research*, vol 11, no 4, pp568–579

Feldman-Savelsberg, P. (2002) 'Is infertility an unrecognized public health and population problem?', in M. Inhorn and F. van Balen (eds) *Infertility Around the Globe: New Thinking on Childlessness, Gender, and Reproductive Technologies*, University of California Press, Berkeley, CA

Frank, A. (2004) *The Renewal of Generosity: Illness, Medicine and How to Live*, University of Chicago Press, London and Chicago

Franklin, S. (1997) *Embodied Progress: A Cultural Account of Assisted Conception*, Routledge, London

Gallagher, C. (2000) 'White like me? Methods, meaning, and manipulation in the field of white studies', in F. Twine and J. Warren (eds) *Racing Research, Researching Race*, New York University Press, London

Gunaratnam, Y. (2003). *Researching 'Race' and Ethnicity: Methods, Knowledge, and Power*, Sage, London

Gunaratnam, Y. (2007) 'Where is the love? Art, aesthetics and research', *Journal of Social Work Practice*, vol 21, no 3, pp271–287

Haraway, D. (1988) 'Situated knowledges: The science question in feminism and the privilege of partial perspective', *Feminist Studies*, vol 14, no 3, pp575–599

Harding, S. (1986) *The Science Question in Feminism*, Cornell University Press, Ithaca, NY

Henry, M. (2007) 'If the shoe fits: Authenticity, authority and agency in feminist diasporic research', *Women's Studies International Forum*, vol 30, no 1, pp70–80

Hollway, W. (2007) 'Afterword', *Infant Observation*, vol 10, no 3, pp331–336

Hosseini, K. (2003) *The Kite Runner*, Bloomsbury, London

Inhorn, M. (2004) 'Privacy, privatization, and the politics of patronage: Ethnographic challenges to penetrating the secret world of Middle Eastern, hospital-based in vitro fertilization', *Social Science & Medicine*, vol 59, no 10, pp2095–2108

Inhorn, M. (2007) 'Masturbation, semen collection and men's IVF experiences: Anxieties in the Muslim world, *Body and Society*, vol 13, no 3, pp37–53

Ipsos Mori (2002) 'Who's Asking? Answers May Depend On It', www.ipsos-mori.com/publications/rmw/whos-asking.shtml, accessed 13 November 2007

James, N. (1989) 'Emotional labour: Skill and work in the social regulation of feelings', *Sociological Review*, vol 37, no 1, pp15–47

Kai, J. and Hedges, C. (1999) 'Minority ethnic community participation in needs assessment and service development in primary care: Perceptions of Pakistani and

Bangladeshi people about psychological distress', *Health Expectations*, vol 2, no 1, pp7–20

Lather, P. and Smithies, C. (1997) *Troubling the Angels: Women Living with HIV/AIDS*, Westview Press, Boulder, CO

Lawrence, E. (1982) 'In the abundance of water, the fool is thirsty: Sociology and black pathology', in CCCS Collective *The Empire Strikes Back*, Hutchinson, London

Leifer, E. (1988) 'Interaction preludes to role setting: Exploratory local action', *American Sociological Review*, vol 53, pp865–878

Letherby, G. (2002) 'Claims and disclaimers: Knowledge, reflexivity and representation in feminist research', *Sociological Research Online*, vol 6, no 4, www.socresonline.org.uk/6/4/letherby.html, accessed 13 November 2007

Liebow, E. (1967) *Tally's Corner: A Study of Negro Streetcorner Men*, Routledge and Kegan Paul, New York and London

Massey, D. (1999) 'Imagining globalisation: Power-geometries of time-space', in A. Brah, M. J. Hickman and M. Mac an Ghaill (eds) *Future Worlds: Migration, Environment and Globalisation*, Macmillan, London

Nast, H. (1994) 'Opening remarks on "Women in the field"', *The Professional Geographer*, vol 46, no 1, pp54–66

Nayak, A. (2006) 'After race: Ethnography, race and post-race theory', *Ethnic and Racial Studies*, vol 29, no 3, pp411–430

Papadopoulos, I. and Lees, S. (2002) 'Developing culturally competent researchers', *Journal of Advanced Nursing*, vol 37, no 3, pp258–264

Phoenix, A. (2001) 'Practising feminist research: The intersection of gender and 'race' in the research process', in K.-K. Bhavani (ed) *Feminism and Race*, Oxford University Press, Oxford. Originally published 1994, in M. Maynard and J. Purvis (eds) *Researching Women's Lives*, Taylor Francis, London

Reay, D. (1996) 'Insider perspectives or stealing the words out of women's mouth: Interpretation in the research process', *Feminist Review*, vol 53, pp57–73

Rhodes, P. (1994) 'Race-of-interviewer effects: A brief comment', *Sociology*, vol 28, no 2, pp547–558

Riessman, C. (1987) 'When gender is not enough: Women interviewing women', *Gender and Society*, vol 1, no 2, pp172–207

Sanghera, G. and Thapar-Bjorkert, S. (2008) 'Methodological dilemmas: Gatekeepers and positionality in Bradford', *Ethnic and Racial Studies*, vol 31, no 1, pp543–562

Schaeffer, N. (1980) 'Evaluating race-of-interviewer effects in a national survey', *Sociological Methods and Research*, vol 8, no 4, pp400–419

Schaeffer, N. and Maynard, D. (2002) 'Standardization and interaction in the survey interview', in J. Gubrium and J. Holstein (eds) *Handbook of Interview Research: Context and Method*, Sage, Thousand Oaks, CA

Schuman, H. and Converse, J. (1971) 'The effects of black and white interviewers on black responses in 1968', *Public Opinion Quarterly*, vol 35, pp44–68

Sewpaul, V. (1999) 'Culture, religion and infertility: A South African perspective', *British Journal of Social Work*, vol 29, no 5, pp741–754

Song, M. and Parker, D. (1995) 'Cultural identity: Disclosing commonality in in-depth interviewing', *Sociology*, vol 29, no 2, pp421–456

Stanfield, J. and Dennis, R. (eds) (1993) *Race and Ethnicity in Research Methods*, Sage, Newbury Park, CA

Thapar-Bjorkert, S. and Henry, M. (2004) 'Reassessing the research relationship: Location, position and power in fieldwork accounts', *International Journal of Social Methodology*, vol 7, no 4, pp363–381

Tuhiwai Smith, L. (1999) *Decolonizing Methodologies: Research and Indigenous Peoples*,

Zed Books and University of Otago Press, London

Twine, F. and Warren, J. (eds) (2000) *Racing Research, Researching Race: Methodological Dilemmas in Critical Race Studies*, New York University Press, London

Urwin, C. (2007) 'Doing infant observation differently? Researching the formation of mothering identities in an inner London borough', *Infant Observation*, vol 10, no 3, pp239–251

Visweswaran, K. (1994) *Fictions of Feminist Ethnography*, University of Minnesota Press, Minneapolis, MN

Part Two

Exploring Infertility, Ethnicity and Culture in National Contexts

6

Commonalities, Differences and Possibilities: Culture and Infertility in British South Asian Communities

Lorraine Culley and Nicky Hudson

Introduction

The ethnic patterning of health in the UK is well established, with studies exploring this in relation to a wide range of diseases, conditions and healthcare experiences (Nazroo, 1997; Aspinall and Jacobson, 2004; Sproston and Mindell, 2006; Bhopal 2007). Despite a strong and long-established regulatory context for fertility treatment, extensive anti-discrimination legislation and the fact that ethnicity monitoring has been mandatory in the National Health Service (NHS) since 1995, there are no published data in the UK on ethnic differences in the prevalence of infertility, or on the ethnic background of those receiving fertility treatment.[1] The national body that regulates fertility treatment, the Human Fertilisation and Embryology Authority (HFEA), has extrapolated from the UK Census data that a quarter of a million people from 'non-white' groups will face infertility; this includes 136,000 individuals of South Asian heritage (HFEA, 2006).[2] However, within the UK there is relatively little empirical research that explores how the experience of infertility is shaped by ethnic identity, particularly in minority communities (see Culley, Chapter 1). It has been suggested by some authors that this failure of research to explore the impact of ethnicity reflects a racist perception that minority women have 'too many' babies (Roberts, 1997; Ceballo, 1999; Katbamna, 2000; Inhorn et al, Chapter 11).

This chapter aims to address this absence in the literature by discussing some of the findings of a research project which explored the experience of involuntary childlessness in British South Asian communities, carried out between 2002 and 2004. The 'ASFERT' study, which was funded by the NHS, was designed both to access 'community' perceptions of infertility and its treatment, and to explore the experiences of British South Asian women and men who were experiencing fertility problems.[3]

The first phase of the study included 14 community-based focus groups with a total of 93 people (27 men and 68 women) of Indian, Pakistani and Bangladeshi origin who were not necessarily themselves experiencing fertility problems. This aspect of the study gave us the opportunity to explore the ways in which infertility is perceived by a range of individuals who are potentially typical of the wider kinship and community networks of childless couples within these communities. To accommodate the religious and cultural requirements of research with South Asian groups (Atkin and Chattoo, 2006; Papadopoulos, 2006), the focus groups were single sex, and also represented different generations (Culley and Hudson, 2007; Culley et al, 2007a). Phase One also included interviews with 21 'key informants' from South Asian communities who provided important contextual data on prevailing social norms. This phase informed the data collection and analysis in the second part of the study which included semi-structured interviews with 37 women and 13 men of South Asian descent who had identified fertility problems and had experienced medicalized infertility treatment.[4] Twenty-three infertility healthcare professionals (medical consultants, fertility nurses and counsellors) were also interviewed, providing data on their perspectives of cultural difference, and information about how fertility clinics respond to difference in everyday practice.

In this chapter we aim to provide an understanding of the ways in which infertility is perceived and experienced within a particular socio-cultural location. We describe community perceptions of childlessness and infertility in British South Asian communities as highly stigmatized, which arises as a consequence of a pronounced pronatalist ideology within British South Asian culture. We also discuss some of the potential implications of this context for those who are experiencing fertility problems. We are arguing, therefore, for a recognition of the significance of culture in understanding the infertility experience and that healthcare providers and policy makers need to take account of cultural difference in their practice. We are also arguing, however, against an essentialist position, whereby ethnic groups are conceptualized as distinct 'cultures' – bounded, internally homogeneous and ordered around essential beliefs and values (Ahmad, 1996) which inevitably determine individual behaviour.

Essentializing culture

We argue that it is important to avoid essentializing South Asian infertile individuals as belonging to an undifferentiated and fixed 'culture' that determines behaviour and experience in a simplistic manner. Health service research is frequently characterized by a cultural essentialism that 'exaggerates the unity of cultures, solidifies differences that are currently more fluid and makes people from other cultures seem more exotic and distinct than they really are' (Phillips, 2007, p14). Cultural essentialism ignores the heterogeneity of cultures and plays down the existence of differentiation and hierarchies of power *within* racialized minority groups (Culley, 2006).

Cultural essentialism not only imagines a false homogeneity of cultures, it also tends to deny agency to members of racially minoritized groups. It represents members of racialized groups as controlled by cultural rules – compelled by their 'culture' to behave in particular ways, often to the alleged detriment of their health. For South Asian women in particular culture is often invoked to construct a passive subject, submissive to male dictates, and family and community expectations. Such stereotypes, however, have been effectively challenged. A growing number of authors recognize minority women as active agents and describe the ways in which women are able to draw on social and cultural resources in order to resist, challenge, adapt (and sometimes re-inscribe) 'traditional' elements of culture (Brah, 1996; Ramji, 2003, 2007; Ahmad, 2006; Phillips, 2007; Hudson, 2008). We argue here, then, for a more nuanced approach to the impact of culture, as something that influences and shapes behaviour but does not necessarily determine it (Kelleher, 1996; Culley, 2008) and an approach that allows for the *possibility* of agency and autonomy.

Our research suggests that within British South Asian 'culture', the perceptions of involuntary childlessness and the experience of infertility are mediated by a number of social signifiers. We argue, first, that cultural understandings of and experiences of infertility are mediated by gender, generation and socioeconomic status. Second, that cultural understandings are fluid and dynamic, and in particular that there are changing social norms around family formation, which affect the experience of infertility. Thirdly, we argue that although cultural norms are an important and inescapable aspect of the social context of the infertile, they are also open to challenge and adaptation and will be negotiated in specific ways by differently placed individuals. Infertile individuals are positioned in a complex nexus of social relations and dynamics.

It is impossible to capture all aspects of this complexity and fluidity within this chapter. However, to illustrate our argument, we discuss key themes from the 'local moral worlds' (Inhorn, 2003) of our participants, which help us to explore the significance of cultural norms and values in the lives of the infertile, and the extent to which these are differentially experienced and actively negotiated. While demonstrating the culturally specific, we also draw out issues that suggest a *commonality* of experience with infertile individuals from white 'majority' communities.

Drawing on this theoretical approach to the conceptualization of culture, the rest of the chapter is organized into three sections. In discussing the data from the focus groups with community members and interviews with infertile couples, we first explore the ideology of pronatalism and the specific cultural inflection of this within South Asian communities. This section illustrates the predominance of pronatalism but also the way in which this is mediated by age and generation and the impact of changing constructions of family life. Next, the chapter discusses some of the significant social consequences of infertility as a highly stigmatized condition in South Asian communities, but also shows the extent to which this is differentially experienced by women and men. Finally, we discuss issues relating to accessing infertility services, and here we illustrate the potential significance and complexity of cultural influences *and* the impact of structural issues on the ability of individuals to access assisted conception. We also argue that some of the difficulties which some South Asian couples experience should be seen as deriving from an inadequate *response* to cultural difference on the part of healthcare providers rather than as emanating from a problematic cultural location.

Pronatalism, culture and generation

To understand attitudes to *in*fertility, it is necessary first to explore attitudes to fertility and the importance of children. Pronatalism can be defined as any attitude or policy that is 'pro-birth', encourages reproduction and exalts parenthood (Miall, 1994; Remennick, 2000). Discourses of pronatalism have a dominant presence in most Western contexts, which means that expectations about childbearing for normative adulthood remain prevalent in all sectors of British society (Throsby, 2004). However, the data from all aspects of our study demonstrate a very intense pronatalist ideology within South Asian communities. Marriage is highly valued and almost universal, and parenthood is perceived as the natural and desired consequence of marriage. Children are seen as an essential aspect of life and parenthood is an almost unquestionable component of adult social identity. As this Bangladeshi woman commented:

> If women can't have children, then they don't have life. Children are life – they are the future.
>
> (Focus group, older Bangladeshi Muslim woman)

Voluntary childlessness was inconceivable for most participants in our study.

> I want to be a mother. All women want to be a mother because we are born as mothers. If I don't have any children then I would feel very bad.
>
> (Focus group, older Bangladeshi Muslim woman)

Our data suggest that child desire in British South Asian communities is multifaceted. Social, economic, religious and emotional reasons for having children were discussed in the focus groups with South Asian participants, who were asked to reflect on the importance of children within their communities (Culley et al, 2004; Culley et al, 2006b). There were, however, some significant generational differences in these perceptions. Older people were more likely to discuss 'social security' reasons for having children, such as providing support for parents in old age. They were also more likely than younger people to point to the importance of children in performing religious rituals for dying parents and in offering prayers for the dead. On the other hand, 'social perpetuity' desires (Inhorn, 2002) were mentioned by all age groups and genders, and the need to continue the family name, lineage or clan was a common theme in all the focus groups.

Interviews with people receiving infertility treatment also confirmed a very strong expectation among their families and the wider community that married couples should produce children. Most reported feeling intense family pressure to produce a child (often from other female members). They also suggested, however, that the negative treatment of childless couples was more likely to come from older members of the community. The desire for a child was not merely seen as a response to external pressure. Individual infertile women in particular also expressed a deeply held personal need to have a child. This is consistent with the findings of most studies of infertile couples (Phoenix et al, 1991; Monach, 1993; Letherby, 1999; Becker, 2000; Ulrich and Weatherall, 2000; Throsby, 2004).

As we have seen, there were differences between generations in terms of the reasons given for the importance of children. Generational diversity was also evident in relation to expectations of the point at which a marriage should produce children and the overall family size, and also in relation to gender preference. Older people made the point that for earlier generations, the expectation was that a child should be born within a short time of marriage, and that the couple would face serious questioning if a pregnancy was not forthcoming. There was, however, a realization in several groups that this was now changing. The increasing acceptance of *delayed* childbearing for married couples was highlighted by a number of participants. The view that young couples would now want to delay parenting until after they had completed their education, or were settled in a house, or in a career, was expressed by a number of participants:

> *It's only because nowadays careers take over rather than in those [days] they didn't and they were just at home and looking after the children and making more children. Nowadays it's more career minded women that are sort of taking over.*
>
> (Focus group, younger Pakistani Muslim woman)

Growing participation in higher education is related to this pattern of later childbearing. Higher education is highly valued in British South Asian communities. There is a very high level of participation among Indian groups, though Pakistani and Bangladeshi women in particular are still considerably less likely to enter higher education than other ethnic groups. Some younger women reported that one of the consequences of going to university was that women were able to defuse some of the pressure to marry and reproduce at an early age and negotiate alternatives with their parents.

The increasing acceptance of delayed childbearing, for more affluent groups at least, also had implications for the management of information about infertility among childless couples (Culley and Hudson, 2006). Infertile women who worked or studied outside of the home were able to use their study or career as an external 'justification' for not having children at that particular point in their lives and were able to give the impression to family and friends that childbearing was simply on hold and thereby avoid difficult or uncomfortable enquiries, as this Indian man said:

> *... so if anyone asks, we just turn around and say, look we wanted to, er, we want to let her to do her study instead of having kids first.*
>
> (Interview, Indian Sikh male)

Early childbearing and larger families are more prevalent in South Asian communities than in the white population (particularly in the Bangladeshi and Pakistani communities), but this picture is changing. A major national study of ethnicity in the UK (Modood et al, 1997) found that Pakistani and Bangladeshi groups had the largest families, with 33 per cent of Pakistani families and 42 per cent of Bangladeshi families having four or more children. Only 4 per cent of white families and 3 per cent of African-Asian families had more than four children while the rate for Indians was 11 per cent. Pakistani and Bangladeshi women are more likely to start families in their 20s than Indian and African-Asian or white women and more likely to continue bearing children later in life (Modood et al, 1997). Figures from the 2001 census confirm a higher family size for most South Asian groups, but also show a slight decline in size for each of the South Asian groups in the period since 1991 (Census, 2001).

Nevertheless, a family with only one child was still regarded by focus group participants as 'deviant', particularly among the older generation. This was reflected in the experience of our interviewees with fertility problems. Over a third of interview participants already had a child and were seeking help to add to their family. Several women reported the negative attitudes of older people to a one-child family. As this Pakistani Muslim woman commented:

> *... but we're still living with our older generation, who has exactly the same ideas because they were able to produce four, five, six kids like, no problem.*
>
> (Interview, Pakistani Muslim woman)

The culturally specific pronatalism we found in our study also encompassed a strongly expressed gender preference, though this was also, to some extent, mediated by generation. A preference for sons in the Indian subcontinent has been widely reported (Bhatia, 1978; Bhatti et al, 1999; Winkvist and Akhtar, 2000; Bharadwaj, 2003) and there was evidence in our study that this is still evident in the UK context. The community-based focus groups and key informants in Phase One reported that the desire for a son was a widely held value. Several participants suggested that not having a male child was a form of 'social' infertility, which could threaten the position of the wife in more 'traditional' families in the same way that an absence of children might. This was explained in terms of South Asian family norms.[5] In many South Asian families, when a daughter marries she leaves her family of origin and enters her husband's family, symbolically if not literally. Sons, however, remain more directly economically and socially obligated to their family of origin, as this Indian Sikh women said:

> *Having boys is a little different. I suppose it's like this... Girls get married and go to their new homes and boys look after their parents.*
>
> (Focus group, older Indian Sikh woman)

The older generation in particular emphasized the importance of sons for religious, cultural and economic reasons, giving examples of how the birth of a boy would be celebrated, while the birth of a daughter sometimes gave rise to the opposite reaction. This was confirmed by some of the younger participants, as this Bangladeshi man observed:

> *All the members of my family's family became upset when my wife gave birth to my daughters and they started back biting about my wife. But when she gave birth to my son they all celebrated.*
>
> (Focus group, Bangladeshi Muslim man)

The birth of a male child in particular was seen to confer social power on women within the marriage and the wider family, as this Indian Sikh woman commented:

> *The more boys you have the more status a female has got, regardless of what she's like or what the children are like.*
>
> (Interview, Indian Sikh woman)

Despite recognizing a strong sense of gender preference within their communities, however, many of the younger participants in the focus groups challenged the idea that boys should be more highly valued than girls, regarding this as an outdated and 'traditional' attitude. Most of our participants who were undergoing fertility treatment, whilst acknowledging that the pressure to have a male child existed, did not themselves express a strong gender preference.

Pronatalism is not of course confined to minority ethnic communities. Studies with white middle class participants in the UK, US and Canada have argued that pronatalism is a key feature of contemporary ideology and that infertility is thus regarded as a deficient, deviant or abnormal condition. Childless women are stigmatized (Whiteford and Gonzalez, 1995; Letherby, 1999) and experience feelings of shame, inadequacy and failure. A rare study of community constructs of infertility carried out in eastern Canada, for example, demonstrates a clear pronatalist ideology. In this study, however, infertility is seen as a largely personal and couple-centred issue (Miall, 1994). Van Balen and Inhorn (2002) have also argued that the possibility of childlessness as a lifestyle choice tends to obscure the visibility of involuntary childlessness. Within British South Asian communities, voluntary childlessness is unthinkable for most people and childlessness for many is experienced as a matter of social concern and social surveillance, as well as a personal 'failing'.

Our findings also demonstrate, however, that although pronounced pronatalist discourses are present in British South Asian communities, these are constructed and experienced differently by different individuals, in different social locations and different sets of family and personal relationships. The interviews demonstrated that reactions to infertility varied between families and were not inevitably negative or unsupportive. Furthermore, while pronatalism remains significant, many young men and women are more likely to be able to choose to 'legitimately' delay childbearing, are more likely to choose to have smaller numbers of children and are more likely to take pleasure in the birth of a daughter, than their parents' generation.

Stigma, culture and gender

The intensity of pronatalism means that childlessness is highly stigmatized and this was a strong theme in our study across all communities and in the individual interviews with those undergoing fertility treatment. Infertility has been described as an invisible, secret stigma whereby the infertile are able to pass as 'normal' (Goffman, 1963). While this may be the case in majority communities (Greil, 1991), in South Asian communities childlessness is highly deviant and highly visible and any attempt at 'passing' is only temporarily successful. The process of stigma, however, is highly mediated by gender.

Infertility is seen almost exclusively as a woman's problem and all participants in the study, young and old, male and female, and across all communities agreed that women are 'blamed' for childlessness. 'It's always the woman's fault' was a very common theme. This, however, is not confined to minority communities in the West. As van Balen and Inhorn (2002) have forcefully argued, it is women, the world over, who bear the burden of shame and social scrutiny for infertility and it is women 'who live the untoward repercussions and social backlash associated with this affliction' (p19).

In our study the implications of childlessness were widely reported as being more severe for women than men. Many examples were given in the focus group and key informant discussions of how childless women might be gossiped about, taunted and in a few extreme cases socially ostracized and ill-treated by families. Several women undergoing treatment reported negative behaviour towards them from their partner's family in particular. Some of the women interviewed felt that childlessness denied them full membership of their husband's family, and that their role in the family would not become secure until they had a child. As this Bangladeshi man commented:

> Some say, 'What is the point of keeping this woman, if she cannot provide children?'
>
> (Focus group, Bangladeshi Muslim man)

Both male and female interviewees confirmed that the woman was always seen as the 'guilty party' responsible for a couple's childlessness, even where the 'fault' was known to lie with the man, as this Pakistani women commented:

> They will never accept that defect lies in their son. They will always blame daughter-in-law. They will taunt her and make her life hell.
>
> (Focus group, older Pakistani Muslim woman)

Even where male factor problems were involved, women were expected not to expose male infertility and allow themselves to 'take the blame', to protect men from stigma.

Very direct and painful questioning of women about their failure to produce a child was reported in our study. As a result, a number of women reported avoiding occasions such as family weddings or other social events where they might be faced with comments and direct questions about their childlessness, as one way in which to at least avoid stigma, if not to escape it entirely:

> *I mean, literally, for a couple of years, I just didn't socialise at all, and didn't go to big functions and parties, when I couldn't have (child's name), because the pressure was so great. People just looked at me, like, you know, she's no use to this community, she'd rather be off somewhere else.*
>
> (Interview, Pakistani Muslim woman)

Our data also confirm the findings of Katbamna (2000) and Inhorn (1996), both of whom reported pressures on women from mothers-in-law or other members of the husband's family in particular. The direct and intrusive questioning of women about their childless status has also been reported in a study of strongly pronatalist Israeli society (Remennick, 2000) and in Reissman's study of South India (2000). However, studies of 'white' infertile women (Imeson and McMurray, 1996), also report insensitive responses from family and friends, although the *possibility* of voluntary childlessness may result in a blurring that obscures the visibility of involuntary childlessness (van Balen and Inhorn, 2002; Throsby, 2004).

Despite the clear finding that the social consequences of infertility are particularly severe for women, our study suggests that men do not entirely escape stigma. Focus group participants suggested that a childless man might have to suffer taunts about his sexuality and lack of 'manhood', since there is a common tendency to link infertility with impotence and virility. None of the men interviewed in this study, however, reported themselves having experienced this kind of reaction to their childlessness. Instead, a number of male interviewees described the pressure that men often come under to divorce their wives and remarry. It was also widely felt in the focus groups that infertility would be considered 'good grounds' for men to consider divorce or (in some communities) the taking of additional wives, and many instances of these 'solutions' to infertility were reported. Several participants made the point that in some families, even where a husband might not wish to divorce his wife, his family might insist on this.

This gendering of the stigma of infertility had the affect of locating the problem with the woman, whilst endowing the male partner with the agency to solve this problem, by the taking of a second or different wife. However, some of the couples we interviewed were clearly able to resist the gendered effects of infertility. Despite the examples in the focus groups of the negative impact of patriarchal social relations, several women spoke of sympathetic and helpful responses from family and from their husband's family.

> *And, from my husband's side of the family – his mother is absolutely brilliant. Yeah, very, very supportive, and she knows what I'm going through.*
>
> (Interview, Pakistani Muslim woman)

Several couples also reported that the experience of infertility had brought them closer together and one-third of women reported that their husbands had been their main source of emotional support.

The above discussion suggests then that the stigma arising from infertility in South Asian culture is intersected by gender, with infertile men and women experiencing this differently. However, our findings also suggest that these responses from others are not universally negative and may be actively resisted and avoided by individuals and couples. Although culture is an important influence on behaviour it is not always directly determinant of experience. Individuals are able to some extent to avoid, resist and adapt existing cultural mores (Kelleher, 1996). The possibility of enacting such strategies, however, although variable, is not entirely contingent. As we saw earlier, the developing trend of delayed childbearing allows some South Asian couples the opportunity to avoid the stigma of infertility, at least temporarily. As with the resistance to pronatalist discourses, the avoidance of stigma was dependent on a woman's particular social positioning. Women from poorer families, those with lower proficiency in English (often, more recent migrants), those who are less likely to have employment outside the home and those from more culturally and religiously 'conservative' families may have fewer opportunities to enact reproductive choice or to escape the consequences of the gendering of infertility.

Accessing treatment

The final part of this chapter discusses how important some aspects of culture might be in influencing treatment seeking for infertility. We do this by first exploring the potential impact of culturally specific understandings of procreation and the causes of infertility, together with culturally and religiously framed attitudes towards treatment options. Following this we demonstrate that while the experience of accessing fertility treatment may be shaped by culture, there are also significant structural enablers and constraints which mediate this effect.

Knowledge of infertility and attitudes to assisted conception

How people explain the causes of ill health, the types of treatment they believe in and the help they seek is culturally variable (Helman, 2001; van Balen, Chapter 2). In Western societies, infertility has become medicalized (Greil, 1991, 2002; Becker and Nachtigall, 1992; Sandelowski and de Lacey, 2002) and the medical model is the dominant cognitive framework in terms of which the experience is interpreted (Sandelowski and de Lacey, 2002). A number of international studies, however, demonstrate that alternative understandings of infertility exist in many societies (Inhorn, 1994; Inhorn and van Balen, 2002; Liamputtong, Chapter 9). For example, infertility may be seen as a punishment from God for wrongdoing in this or a previous life, the result of bad luck, evil spirits or a curse (Inhorn, 1994).

Data from all aspects of our study suggest that treatment seeking is highly nuanced and can be simultaneously informed by a number of models of understanding. A small minority of (mainly older) participants suggested that infertility was related to divine retribution or other spiritual or external influences. A focus group of older Sikh men suggested that infertility might be the outcome of bad deeds in a former existence. A number of Muslim focus group participants made the point that Allah was ultimately the giver of children. Although, even where the hand of God was proposed, it was widely accepted in the Muslim groups that seeking medical treatment to correct infertility was also a religious duty.

Among the community groups, knowledge of specific medical treatments for infertility was limited. Some of the health professionals we interviewed suggested that South Asian patients on the whole had less knowledge of fertility than most 'white' patients, although this was clearly influenced by educational levels and what healthcare providers referred to as degrees of 'Westernization'.[6]

The majority of participants in our focus groups were aware of in vitro fertilization ('test tube baby') and felt that IVF would be considered an acceptable treatment for infertility, providing the resultant child was biologically related to the couple. The use of third party assisted conception, however, was universally regarded as socially unacceptable. It was felt that using donor gametes would be a last resort for childless couples and that if they were 'desperate' enough to take this option, then they would most certainly not disclose this to others. In considering treatment options, religion was introduced into some of the discussions by the participants as a lens through which to view certain aspects of infertility treatment. This was more commonly the case within the Muslim groups and here too religious objections to the use of donated sperm were most strongly expressed:

> *It doesn't matter if it's a man or a woman, it's not allowed. Say for instance I go and get pregnant, mixing with a different male, in different ways, having their sperm, that is* haram, *that is sinful in our Islam and in the eyes of the community.*
>
> (Focus group, older Bangladeshi Muslim woman)

The use of donated eggs, however, appeared to be marginally more acceptable.[7] It was clear, then, that IVF treatment had become acknowledged as a socially legitimate intervention to overcome infertility. This was happening in different ways in different communities, with some participants stressing the coexistence of solutions located in biomedicine, religion and 'alternative' therapies.

'Alternative' approaches to infertility treatment

Given the culturally informed discourses of the causes of, and solutions to, infertility, focus group participants and key informants spoke of a range of non-medical therapies and treatments that might be sought out by infertile women. Childless women might visit healers, be given amulets with special prayers, undertake fasts or pilgrimages and/or pray to specific gods. The older members, in particular, discussed the use of various herbal preparations and special foods that could increase fertility. In most cases, however, all such activity was seen as complementary to medical help, rather than as an alternative source of treatment, at least in the UK context. The individual interviews with infertile couples confirmed that around 15 per cent had used 'alternative' or traditional remedies including herbal preparations, acupuncture, homeopathy and massage. This is a similar level to that found in a Dutch survey of infertility behaviour (van Balen et al, 1995, cited in van Balen and Inhorn, 2002, p10).

Within South Asian communities, as within other social groups, treatment seeking strategies are highly nuanced – influenced by culture, but not determined by it in all cases. There are few 'cultural' barriers to treatment in the sense of health beliefs or attitudes to biomedicine. However, for some of those who attempt to access infertility treatment, the institutional *response* to cultural difference can create particular difficulties. One example of this can be seen in the case of those individuals who are less proficient in the English language.

Communication support and institutional discrimination

In infertility management, treatment regimes are often very complicated; there are potentially dangerous side-effects and difficult decisions often have to made, such as those concerning the use of donated gametes and the preservation and disposal of embryos. Good communication is therefore an essential component of high quality care. Differences of language between patient and provider in the UK healthcare system, however, often present significant difficulties for patients (Bradby, 2001; Szczepura et al, 2005) and the failure to provide adequate systems of language support is regarded by many as a prime example of institutional discrimination.

Although most South Asian individuals of childbearing age speak good English, there is a significant minority who do not, especially older women in the Bangladeshi and Pakistani communities (Modood et al, 1997) and those spouses who are more recent migrants to the UK. The need for high quality interpreting services has long been recognized as essential to understand and meet the needs of people who do not speak the majority language. Communication problems mean that patients are less likely to consult about ill health in the first place, the development of trust between patient and practitioner is hindered, proper informed consent is difficult to achieve, a lack of compliance with medical regimes is possible and concordance is unlikely (Rhodes and Nocon, 2003).

In our study, one-quarter of those actively involved in infertility treatment did not speak English and this undoubtedly created considerable difficulties for them. Although clinics provided extensive written information about infertility and its management, this was only available in English; no translated consent forms or details of treatment options, outcomes or potential side-effects were available.[8] Healthcare providers also reported a very limited use of official, trained interpreters. *Informal* interpreters including relatives of patients and staff from other areas of the hospital were commonly used. This is particularly inappropriate in infertility consultations which often contain reference to sexual practices, intimate body parts and other highly personal details. Some nurses reported resorting to mime and gesture to communicate.

Several of our female interview participants spoke of considerable stress when they were required to interpret complex, personal and sensitive issues for their partners at the same time as assimilating often emotionally difficult issues themselves. When questioned directly, some healthcare providers also expressed concern where partners or other informal interpreters were used in consultations.

Several raised serious doubts about the veracity, completeness and accuracy of the translations, especially in the case of male partners interpreting for their wives.

> *I know I've been in this situation a few times, really, where you think – I'm not convinced that the lady is aware exactly what's going on here. Which, from a professional point of view, is not good is it, really?*
>
> (Infertility nurse)

The important point here is that the issue should not be perceived as a 'problem' emanating from the cultural distinctiveness of patients who are not proficient in English. The problem is one that is created by an inadequate *response* to cultural difference, which could be regarded as an example of institutional discrimination (see Atkin, Chapter 3). It certainly represents a significant barrier to good quality treatment for those who are not proficient in English and is a clear instance where 'culture' matters, in this case because of the failure of institutions to recognize and respond appropriately to cultural difference.

Ethnicity and socio-economic status

The tendency within cultural essentialist approaches to healthcare is to regard culture as a significant constraint on access to treatment. In contrast to debates about access for majority populations, in the case of minoritized groups, *cultural* constraints loom larger than other influences such as social class. This is in line with the tendency to represent such groups as uniquely driven by their cultural values and traditions. However, our study again suggests a more complex understanding is required. Where studies have sought to explore access to healthcare in a wider context, it has been shown that effective access may be less influenced by cultural identity than more general educational levels and socio-economic status (Hennink et al, 1998; Rhodes et al, 2003).

The impact of socio-economic status and its intersection with ethnic identity were also important in our study. In the UK, while initial investigations and some forms of treatment are available within the publicly funded NHS sector, provision of IVF and other complex procedures is very limited.[9] There is no *national* entitlement to treatment within the NHS and, in effect, availability of treatment is a 'postcode lottery'. Some local commissioning agencies (Primary Care Trusts) do not fund any IVF, others offer one, two or three cycles of NHS-funded treatment (BioNews, 2007). There are also variable and increasingly restrictive eligibility criteria in operation for publicly funded treatment, in relation to age, the existence of any previous children, marital status and so on. This creates extreme inequity of provision between people according to where they live (Harrison, 2005). At the present time in the UK around 75 per cent of IVF cycles are funded by patients (HFEA, 2008). A number of our participants described the tremendous financial burden that paying for infertility treatment presented, especially for those who already had limited economic resources.

The implications for patients of the inconsistent and inadequate funding of infertility treatment in England and Wales are clear, regardless of ethnic identity. However, most minority ethnic groups in the UK are economically disadvantaged compared with the general population. This is particularly acute for the Pakistani and Bangladeshi communities, where over 60 per cent of households have incomes below the poverty line. Even the more 'prosperous' minority groups such as the East African Asian group are worse off than their white equivalents in income terms (Platt, 2002; Nazroo, 2006). For poorer couples, then, treatment options will be limited by ability to pay. So, for example, in the case of male factor sub-fertility, donor insemination treatment (which is less effective) will be provided by the NHS but IVF/ICSI (intracytoplasmic sperm injection, using partner sperm) might only be available in the independent sector.

The impact of socio-economic status and access to financial resources therefore has the potential to be a significant variable in understanding the experience of infertility in British South Asian communities (as is the case for minority groups in the US, see Inhorn et al, Chapter 11). As we have shown in this discussion, although cultural and religious understandings of infertility and culturally nuanced attitudes to infertility treatments remain important influences for some, these issues are clearly intersected with structural and economic factors that can facilitate or impede a couples' access to treatment.

Conclusion

Our discussion confirms the importance of seeing infertility, not just as a medical or even a psychological phenomenon, but as socially constructed and negotiated. The concept of infertility as a socio-culturally constructed experience implies that wider social perceptions of infertility must be taken into account by those who fund and provide infertility services.

In presenting our study, we have also argued that it is important to avoid essentializing culture, and remain alert to the similarities in the experiences of majority and minority populations (Atkin, 2004; Culley et al, 2006b). Most societies display a pronatalist ideology and there is evidence that infertility can have significant consequences for women in many communities. Many of the needs and concerns articulated by our infertile South Asian participants can arise for people from all ethnic backgrounds. Although we have not discussed in detail the treatment experiences of our participants, their concerns about financing treatment, about a lack of information and about inadequate emotional support were not dramatically different from those reported by 'white' patients (see Culley et al, 2006a).

Nevertheless, in the accounts of our participants we can see a changing but still distinct cultural particularity in perceptions of parenthood and infertility. In British South Asian communities, specific cultural norms and values define a particular cultural location, which may have a specific and profound impact on people experiencing fertility problems (albeit one that is currently not widely recognized or responded to adequately by healthcare providers). We have described a particularly strong pronatalism and its consequences for infertility as a highly stigmatized condition.

Retreating from cultural essentialist discourses does not mean denying that people are 'cultural beings'. As Phillips argues, 'Everyone is shaped in some ways that we recognise and others of which we remain largely unconscious, by the norms and practices through which we have become the people we currently are' (2007, p52). However, this is different to saying that people are from 'a particular culture', conceived of as a unity of beliefs, practices and ways of understandings, which will inevitably explain what individuals 'do'. Cultural and religious meanings are best seen as broad guidelines for action, internally contested, open to fluid interpretation, and highly dynamic rather than prescriptive codes (Kelleher, 1996; Karlsen and Nazroo, 2002; Bradby, 2003).

Although the influence of cultural identity is evident in our study, we can also see the significance for individuals of intersecting statuses which mediate culture. Age, gender and socio-economic status have been highlighted here, but other facets of identity, such as migration status, are also implicated in the individual experience of infertility and its treatment. Discourses of infertility are constructed and negotiated differently by different generations, different genders and different social classes *and* are dynamic processes. Furthermore, although parenthood remains an essential state for normative adult life, the specific ways in which it is enacted are open to contestation and negotiation. We have argued for an approach to cultural difference that acknowledges the relevance of culture without making it an inevitable and invariable determinant of action, and which allows for internal differentiation in the experience of culture and its effects.

We have also demonstrated that when minority ethnic couples experience fertility problems, cultural and religious issues may influence help-seeking behaviour, but structural issues, such as institutional discrimination and the inadequate provision of publicly funded IVF, may be as important as cultural norms and values in determining whether the needs of infertile people are met.

Notes

1   According to the Human Fertilisation and Embryology Authority an estimated one in seven couples have difficulty conceiving and infertility is the commonest reason for women aged 20–45 to visit their general practitioner (GP) after pregnancy itself. In the UK, in 2005 36,623 patients underwent IVF treatment and a further 2624 patients underwent donor insemination, resulting in a total of almost 12,000 babies (around 1 per cent of all births). Only clinics licensed by the HFEA and subject to their Code of Practice can carry out treatments such as intrauterine insemination (IUI), IVF and ICSI, and can store gametes or embryos (van den Akker, 2002; HFEA, 2007).

2   Minority ethnic groups in the UK are most commonly classified according to the methods used by the population census, which includes an ethnicity question. According to the 2001 census, 92 per cent of the UK population defined themselves as white, with 7.9 per cent (4.6 million people) classifying themselves in a non-white category. Of these, over 2.3 million people (around 4 per cent of the population) described their ethnic origin as Asian or Asian British, which in the UK context refers to those with family origins in India, Pakistan or Bangladesh. A further 1 per cent of the population gave their ethnic origin as Black African, Black Caribbean or Black Other.

3   Although they share common geographical origins in the subcontinent, it is recognized that the term South Asian frequently obscures the complexity of the communities it is seeking to describe (Jayaram, 2004; Ramji, 2006). We attempted to take this into account in our delineation of 'communities' (Gujarati Hindu, Punjabi Sikh, Bangladeshi Muslim and Pakistani Muslim, see Culley et al, 2004), but we recognize the problematic nature of even these generalized categories. However, on many key issues, there was a high degree of consensus between the different ethnic groups identified, although religious issues were more commonly highlighted by the Muslim groups.

4   The methodology is described in detail elsewhere (Culley et al, 2007a). The community-based focus groups were held in the preferred language of participants, tape-recorded, translated and transcribed verbatim. The participants for the second phase were recruited via infertility clinics in three English cities. Of these, 32 gave their ethnicity as Indian, 11 as Pakistani and four as Bangladeshi, giving a broad representation of the percentages of these groups in the UK population. Three participants gave their ethnic origin as 'other', which included two who described themselves as Sri Lankan and one 'Asian other'.

5   In the Indian subcontinent the wide availability of prenatal sex determination, even in rural areas, is leading to a growing sex imbalance in the population in many parts of the country. An estimated 10 million female foetuses have been aborted as a consequence in the decade 2005–2006 (Jha et al, 2006). As Mallik (2002) has argued, to many families, new technologies for preconception and antenatal sex selection are seen today as cost-effective tools for patrilineal kinship-building by eliminating the expense of rearing unwanted daughters. This does not necessarily imply the lack of women's agency in this process, as some women perceive this 'choice' as empowering, allowing them to negotiate their place in the family and community and to gain status (Mallik, 2002).

6   Although infertility is not widely discussed in public arenas, there was, in fact, no reluctance to discuss this in the focus groups and there was indeed a great deal of interest in learning more about causes and treatments (Culley et al, 2007b).

7   This finding was confirmed in our study of third party assisted conception, funded by the ESRC (RES-160-25-0044) which explored specifically the perceptions of gamete donation among British South Asian communities (Culley et al, 2006a).

8   As part of this study, the authors produced an information resource: *Trying for a Baby*, which is available in leaflet and CD format, in English, Gujarati, Bengali, Urdu and Punjabi. The leaflet can be downloaded free from: www.dmu.ac.uk/faculties/hls/nursingmidwifery/resource/index.jsp

9   The National Institute for Health and Clinical Excellence (NICE) Clinical Guideline 11 (2004) provides a more reliable framework for the provision of infertility services, suggesting that three funded cycles of IVF should be provided for eligible couples. However, this has yet to be implemented across the NHS.

References

Ahmad, F. (2006) 'The scandal of "arranged marriages" and the pathologisation of BrAsian families', in N. Ali, V. S. Kalra and S. Sayyid (eds) *A Postcolonial People: South Asians in Britain*, C. Hurst & Co, London

Ahmad, W. I. U. (1996) 'The trouble with culture', in D. Kelleher and S. M. Hillier (eds) *Researching Cultural Differences in Health*, Routledge, London

Akker, O. van den (2002) *The Complete Guide to Infertility: Diagnosis, Treatment, Options*, Free Association Books, London

Aspinall, P. and Jacobson, B. (2004) *Ethnic Disparities in Health and Health Care*, Department of Health, London

Atkin, K. (2004) 'Institutional racism, policy and practice', in S. Ali and K. Atkin (eds) *Primary Healthcare and South Asian Populations: Meeting the Challenges*, Radcliffe, Abingdon

Atkin, K. and Chattoo, S. (2006) 'Approaches to conducting qualitative research in ethnically diverse populations', in J. Y. Nazroo (ed) *Health and Social Research in Multiethnic Societies*, Routledge, London, pp95–115

Balen, F. van and Inhorn, M. (2002) 'Introduction. Interpreting infertility: A view from the social sciences', in M. Inhorn and F. van Balen (eds) *Infertility around the Globe: New Thinking on Childlessness, Gender and Reproductive Technologies*, University of California Press, Berkeley, CA

Becker, G. (2000) *The Elusive Embryo: How Women and Men Approach New Reproductive Technologies*, University of California Press, London

Becker, G. and Nachtigall, R. D. (1992) 'Eager for medicalisation: The social production of infertility as a disease', *Sociology of Health and Illness*, vol 14, no 4, pp456–471

Bharadwaj, A. (2003) 'Why adoption is not an option in India: The visibility of infertility, the secrecy of donor insemination, and other cultural complexities', *Social Science and Medicine*, vol 56, no 9, pp1867–1880

Bhatia, J. C. (1978) 'Ideal number and sex preference of children in India', *Journal of Family Welfare*, vol 24, pp3–16

Bhatti, L. I., Fikree, F. F. and Khan, A. (1999) 'The quest of infertile women in squatter settlements of Karachi, Pakistan: A qualitative study', *Social Science & Medicine*, vol 49, no 5, pp637–649

Bhopal, R. S. (2007) *Ethnicity, Race and Health in Multicultural Societies*, Oxford University Press, Oxford

BioNews (2007) 'Fertility treatment in UK remains inadequate', www.bionews.org.uk/new.lasso?storyid=3539, accessed 18 April 2008

Bradby, H. (2001) 'Communication, interpretation and translation', in L. Culley and S. Dyson (eds) *Ethnicity and Nursing Practice*, Palgrave, Basingstoke

Bradby, H. (2003) 'Describing ethnicity in health research', *Ethnicity & Health*, vol 8, no 1, pp5–13

Brah, A. (1996) *Cartographies of Diaspora: Contesting Identities*, Routledge, London

Ceballo, R. (1999) '"The only black woman walking the face of the earth who cannot have a baby": Two women's stories', in M. Romero and A. J. Stewart (eds) *Women's Untold Stories: Breaking Silence, Talking Back, Voicing Complexity*, Routledge, New York

Census (2001) *Ethnicity and Religion*, www.statistics.gov.uk/census2001/profiles/rank/rank_ethnic.asp, accessed 13 June 2008

Culley, L. (2006) 'Transcending transculturalism? Race, ethnicity and health-care', *Nursing Inquiry*, vol 13, no 2, pp144–153

Culley, L. (2008) 'Cultural diversity and nursing practice', *Journal of Research in Nursing*, vol 13, no 2, pp86–88

Culley, L. and Hudson, N. (2006) 'Diverse bodies and disrupted reproduction: Infertility and minority ethnic communities in the UK', *The International Journal of Diversity in Organisations, Communities and Nations*, vol 5, no 2, pp117–126

Culley, L. and Hudson, N. (2007) 'Public understandings of science: British South Asian men's perceptions of third party assisted conception', *The International Journal of Interdisciplinary Social Sciences*, vol 2, no 4, pp79–86

Culley, L., Rapport, F., Johnson, M., Katbamna, S. and Hudson, N. (2004) 'Improving policy and practice: A study of the provision of infertility services to South Asian communities', Report to Dept. of Health, De Montfort University, Leicester

Culley, L., Hudson, N., Rapport, F., Johnson, M. and Bharadwaj, A. (2006a) 'Public perceptions of gamete donation in British South Asian communities', Report on ESRC study, De Montfort University, Leicester

Culley, L. A., Hudson, N., Rapport, F. L., Katbamna, S. and Johnson, M. R. D. (2006b) 'British South Asian communities and infertility services', *Human Fertility*, vol 9, no 1, pp37–45

Culley, L., Hudson, N. and Rapport, F. (2007a) 'Using focus groups with minority ethnic communities: Researching infertility in British South Asian communities', *Qualitative Health Research*, vol 17, no 1, pp102–112

Culley, L., Hudson, N., Rapport, F., Katbamna, S. and Johnson, M. (2007b) '"I know about one treatment where they keep the egg somewhere": British South Asian community understandings of infertility and its treatment', *Diversity in Health and Social Care*, vol 4, no 2, pp113–121

Goffman, E. (1963) *Stigma: Notes on the Management of Spoiled Identity*, Prentice-Hall, Englewood Cliffs, NJ

Greil, A. L. (1991) *Not Yet Pregnant: Infertile Couples in Contemporary America*, Rutgers University Press, London

Greil, A. L. (2002) 'Infertile bodies: Medicalization, metaphor and agency', in M. Inhorn and F. van Balen (eds) *Infertility around the Globe: New Thinking on Childlessness, Gender and Reproductive Technologies*, University of California Press, Berkeley, CA

Harrison, L. (2005) *A Survey Measuring the Impact of NICE Guideline 11: Fertility: Assessment and Treatment for People with Fertility Problems*, National Institute for Clinical Excellence, London

Helman, C. G. (2001) *Culture, Health and Illness* (4th edition) Arnold, London

Hennink, M., Cooper, P. and Diamond, I. (1998) 'Asian women's use of family planning services', *The British Journal of Family Planning*, vol 24, pp43–52

HFEA (2006) 'Communities urged to speak out to ensure they get the fertility services they need', press release, www.hfea.gov.uk/en/1445.html, accessed 20 March 2008

HFEA (2007) *HFEA Code of Practice 7th Edition*, http://cop.hfea.gov.uk/cop/pdf/COPv2.pdf, accessed 26 April 2008

HFEA (2008) *Facts and Figures*, www.hfea.gov.uk/en/406.html, accessed 18 April 2008

Hudson, N. (2008) 'Infertility in British South Asian communities: Negotiating the community and the clinic', unpublished PhD thesis, De Montfort University, Leicester

Imeson, M. and McMurray, A. (1996) 'Couples' experiences of infertility: A phenomenological study', *Journal of Advanced Nursing*, vol 24, no 5, pp1014–1022

Inhorn, M. (1994) *Quest for Conception: Gender, Infertility, and Egyptian Medical Traditions*, University of Pennsylvania Press, Philadelphia, PA

Inhorn, M. (1996) *Infertility and Patriarchy: The Cultural Politics of Gender and Family Life in Egypt*, University of Pennsylvania Press, Philadelphia, PA

Inhorn, M. (2002) 'The "local" confronts the "global": Infertile bodies and new reproductive technologies in Egypt', in M. Inhorn and F. van Balen (eds) *Infertility around the Globe: New Thinking on Childlessness, Gender and Reproductive Technologies*, University of California Press, Berkeley, CA

Inhorn, M. (2003) *Local Babies, Global Science: Gender, Religion and In Vitro Fertilization in Egypt*, Routledge, London

Inhorn, M. C. and van Balen, F. (eds) (2002) *Infertility around the Globe: New Thinking on Childlessness, Gender, and Reproductive Technologies*, University of California Press, Berkeley, CA

Jayaram, N. (ed) (2004) *The Indian Diaspora: Dynamics of Migration*, Sage Publications, London

Jha, P., Kumar, R., Vasa, P., Dhingra, N., Thiruchelvam, D. and Moineddin, R. (2006) 'Low male-to-female sex ratio of children born in India: National survey of 1.1 million households', *The Lancet*, vol 367, no 9506, pp211–218

Karlsen, S. and Nazroo, J.Y. (2002) 'Agency and structure: The impact of ethnic identity and racism on the health of ethnic minority people', *Sociology of Health & Illness*, vol 24, no 1, pp1–20

Katbamna, S. (2000) *Race and Childbirth*, Open University Press, Buckingham

Kelleher, D. (1996) 'A defence of the terms of "ethnicity" and "culture"', in D. Kelleher and S. M. Hillier (eds) *Researching Cultural Differences in Health*, Routledge, London, pp69–90

Letherby, G. (1999) 'Other than mother and mothers as others: The experience of motherhood and non-motherhood in relation to "infertility" and "involuntary childlessness"', *Women's Studies International Forum*, vol 22, no 3, pp359–372

Mallik, R. (2002) 'Sex selection: A gender-based preference for a pregnancy', *Reproductive Health Matters*, vol 10, no 19, pp189–190

Miall, C. E. (1994) 'Community constructs of involuntary childlessness: Sympathy, stigma, and social support', *The Canadian Review of Sociology and Anthropology*, vol 31, no 4, pp392–421

Modood, T., Berthoud, R., Lakey, J. Nazroo, J. and Smith, P. (1997) *Ethnic Minorities in Britain: Diversity and Disadvantage*, Policy Studies Institute, London

Monach, J. H. (1993) *Childless, No Choice: The Experience of Involuntary Childlessness*, Routledge, London

Nazroo, J. Y. (1997) *The Health of Britain's Ethnic Minorities: Findings from a National Survey*, Policy Studies Institute, London

Nazroo, J.Y. (2006) 'Demography of multicultural Britain', in J.Y. Nazroo (ed) *Health and Social Research in Multiethnic Societies*, Routledge, London, pp1–19

NICE (2004) *Fertility: Assessment and Treatment for People with Fertility Problems*, Clinical Guideline 11, www.nice.org.uk/nicemedia/pdf/CG011niceguideline.pdf, accessed 18 April 2008

Papadopoulos, I. (2006) 'Culturally competent research: A model for its development', in J. Y. Nazroo (ed) *Health and Social Research in Multiethnic Societies*, Routledge, London

Phillips, A. (2007) *Multiculturalism Without Culture*, Princeton University Press, Princeton, NJ

Phoenix, A., Woollett, A. and Lloyd, E. (1991) *Motherhood: Meanings, Practices and Ideologies*, Sage Publications, London

Platt, L. (2002) *Parallel Lives?: Poverty Among Ethnic Minority Groups in Britain*, Child Poverty Action Group (CPAG), London

Ramji, H. (2003) 'Engendering diasporic identities', in N. Purwar and P. Raghuram (eds) *South Asian Women in the Diaspora*, Berg, Oxford

Ramji, H. (2006) 'Journeys of difference: The use of migratory narratives among British Hindu Gujaratis', *Ethnic and Racial Studies*, vol 29, no 4, pp702–724

Ramji, H. (2007) 'Dynamics of religion and gender amongst young British Muslims', *Sociology*, vol 41, no 6, pp1171–1189

Reissman, C. K. (2000) 'Stigma and everyday practices: Childless women in south India', *Gender & Society*, vol 14, no 1, pp111–135

Remennick, L. (2000) 'Childless in the land of imperative motherhood: Stigma and coping among infertile Israeli women', *Sex Roles*, vol 43, no 11, pp821–841

Rhodes, P. and Nocon, A. (2003) 'A problem of communication? Diabetes care among Bangladeshi people in Bradford', *Health & Social Care in the Community*, vol 11, no 1, pp45–54

Rhodes, P., Nocon, A. and Wright, J. (2003) 'Access to diabetes services: The experiences of Bangladeshi people in Bradford, UK', *Ethnicity and Health*, vol 8, no 3, pp171–188

Roberts, D. (1997) *Killing the Black Body: Race, Reproduction, and the Meaning of Liberty*, Pantheon Books, New York

Sandelowski, M. and de Lacey, S. (2002) 'The uses of "disease": Infertility as a rhetorical vehicle', in M. Inhorn and F. van Balen (eds) *Infertility around the Globe: New Thinking on Childlessness, Gender and Reproductive Technologies*, University of California Press, Berkeley, CA

Sproston, K. and Mindell, J. (eds) (2006) *Health Survey for England 2004: The Health of Minority Ethnic Groups, Volume 1*, The Information Centre, Leeds

Szczepura, A., Johnson, M., Gumber, A., Jones, K., Clay, D. and Shaw, A. (2005) 'An overview of the research evidence on ethnicity and communication in healthcare', Report to the Department of Health, University of Warwick (CEEHD), Coventry

Throsby, K. (2004) *When IVF Fails: Feminism, Infertility and the Negotiation of Normality*, Palgrave Macmillan, Basingstoke

Ulrich, M. and Weatherall, A. (2000) 'Motherhood and infertility: Viewing motherhood through the lens of infertility', *Feminism & Psychology*, vol 10, no 3, pp323–326

Whiteford, L. M. and Gonzalez, L. (1995) 'Stigma: The hidden burden of infertility', *Social Science & Medicine*, vol 40, no 1, pp27–36

Winkvist, A. and Akhtar, H. Z. (2000) 'God should give daughters to rich families only: Attitudes towards childbearing among low-income women in Punjab, Pakistan', *Social Science & Medicine*, vol 51, no 1, pp73–81

7

'Anything to Become a Mother': Migrant Turkish Women's Experiences of Involuntary Childlessness and Assisted Reproductive Technologies in London

Zeynep Gürtin-Broadbent

According to the 2001 census, ethnic minorities constitute 7.9 per cent of the British population. However, as there are varied distribution and settlement patterns across the country, the localized reality can be appreciably different. For example, in London, where 45 per cent of all Britain's ethnic minorities live, they comprise a significantly larger 29 per cent of the capital's residents (UK Census, 2001). The roots of these minority populations stretch to all corners of the globe, as they narrate vastly varied causes and circumstances of dislocation. Though these populations may at times be collectively subsumed under the 'ethnic minorities' label in today's multicultural, or multi-ethnic, Britain, their members hold little (besides a presumed difference from the majority) in common.

Although a growing body of research into health and social welfare has recently been commissioned and conducted, where ethnicity has been a central concern (Nazroo, 2006), there is still little known about Britain's smaller ethnic minorities. Community-specific characteristics, and intra-group heterogeneity, particularly with reference to healthcare and attitudes, may become obscured under broad and blunt categorizations, most commonly utilized for hospital data collection and policy documents. The broad questions of how to make sense of such diversity, as well as the practicalities of ethnically sensitive research, have been discussed elsewhere in this volume (particularly in Chapters 2 and 3); I wish here to disentangle the Turkish community's experiences from the 'minority' mass, and to highlight some of the particular challenges of such a project.

Below, I have drawn on ethnographic research to provide a succinct portrayal of the Turkish community in London, to facilitate a better contextualization of their views about kinship and family life. Using data from in-depth interviews, with nine migrant Turkish women, I have tried to depict in detail the unique or relevant aspects of their experience with regard to involuntary childlessness and assisted reproduction. The women were recruited into this study through various community channels, and using snowball sampling. They had all undergone a period of involuntary childlessness and had pursued some form of biomedical treatment for infertility within the last five years. All of them lived in London, but regularly accessed Turkish media, were married to Turkish men, spoke Turkish at home, and classified themselves as having predominantly Turkish social networks, although they also had differences regarding migration history and socio-economic status. Three of the women were childless at the time of the interview, four had one child, and two had two children; all, except one, were conceived with assisted reproduction. Since the sample is very small, and the community to which I refer heterogeneous in various ways, it is not my intention to draw broad generalizations or to provide the narrative of a 'typical' experience. Rather, I aim here, with the benefit of rich qualitative data, to portray the particular context of this community, and to provide a basis from which more sophisticated understandings may be developed.

The 'Turkish community'

Not counting the illegal immigrants (who de facto fall outside statistics), there are currently more than 3 million Turks, and descendants of Turks, in Western Europe. They constitute the largest non-European immigrant group within the EU (Manço, 2004). There are notable Turkish communities in Germany, France, the Netherlands and Austria, as well as in Belgium, Denmark, Switzerland, Sweden and the UK. According to the 1991 census, the population of the Turkish community in Britain is 156,000 (UK Census, 1991). However, this is likely to be a significant under-estimate (Turkish Embassy, 2008), since it excludes many naturalized British citizens, as well as illegal immigrants. The Turkish community is an acutely difficult group to define, since they 'do not occupy a clear position in the white/non-white divide on which current understandings of ethnic minorities are based' (Enneli et al, 2005). Moreover, because of Turkey's geographical position, and its strong ethos of 'secular Islam', the Turkish diaspora cannot be made to fit entirely comfortably under 'Middle Eastern', 'Muslim' or 'Mediterranean' umbrellas.

There are three main groups within what I am referring to as 'the Turkish community': Turks from Northern Cyprus; Turks from Turkey; and Kurds from Turkey. These groups have found their way to Britain with different patterns of migration and have arrived in different waves, beginning at around 1945 and continuing to the present day. The current diaspora is made up of an amalgam of people of foreign birth, who may have arrived in the UK for a variety of reasons, including economic and marriage migrants (of both legal and illegal varieties), and political refugees, as well as British-born descendants of all of these. Most Turks have settled in London, especially Lewisham, Lambeth and Southwark in the south, and Enfield, Haringey, Islington and Hackney in the north, forming pockets of dense spatial concentration.

A 'Little Turkey', within which it is possible to find everything one might need, can be visited at certain points along Green Lanes, and in particular along the Kingsland Road from Dalston to Stoke Newington, and offers London's Turks the option to maintain an almost exclusively 'Turkish' existence. Here, the vibrant local community has formed a self-sufficient enclave, where restaurants, hairdressers, banks, *halal* butchers, jewellers and florists with Turkish signs and assistants cater to their predominantly Turkish clientele. Indeed, it is impossible to fail to notice the plethora of Turkish stores selling a variety of Turkish branded goods, from yoghurt to bridal gowns, and music CDs to *çaydanlık* (traditional two-tier teapots). In addition, there are mosques for prayer and religious gatherings, community centres for social support and advice, and social clubs and *kahves* affording insulated (predominantly male) Turkish spaces, within which traditional forms of social recreation may be maintained. There is even a local cinema showing recent Turkish films, a theatre that sometimes puts on Turkish productions, and various venues where touring Turkish musicians give regular concerts.

The community has immediate access to Turkish popular culture and current events via the media. Newsagents stock large varieties of Turkish newspapers and magazines, and satellite dishes broadcast multiple Turkish TV channels directly into London homes. The Turkish football league, soap operas and 'Paparazzi' shows about celebrities are avidly followed in London, as they are in Turkey. In addition there is a radio station and local newspapers, such as *Toplum Postası*, produced by and specifically for London's Turks. Thus the migrants maintain close contact both within their community, and with events, trends and people back 'home'.

Although there is clearly some degree of heterogeneity within this group regarding origin and migrational context, there is nevertheless a coherent group identity, evidenced through cultural cohesion, religious unity, social clustering, the maintenance of intimate transnational links with Turkey, and continued use of the Turkish language. It is then reasonable to expect that this community may have group-specific attitudes towards, and experiences of, infertility and fertility treatment, that is divergent in some ways from a generic 'British' experience, or from the experience of other minorities. Whether this closely replicates Turkish experiences (as observed in the general Turkish population), or hybridizes Turkish and British experiences in some way, will depend on the extent to which 'home' and 'host' cultural frameworks have been maintained, mixed, or reconfigured through processes of acculturation and adaptation (Berry, 1997). Whilst the views expressed below, with regard to the meaning and importance of the family, reasons for desiring children, reactions to infertility, and feelings about possible treatment options, clearly derive from the interviewed individuals' particular trajectories, they nevertheless highlight trends and similarities that are attributable to a shared culture.

Families and the value of children

All societies value the birth of children. However, the reasons why children are desired, and the corresponding pressures on men and women to reproduce may differ significantly between cultures (van Balen and Bos, 2004). In Turkish society, family remains one of the most important social institutions. Entering a heterosexual marriage, and bearing children within that relationship, are central to the normative and presumed life narrative for all individuals (Kâğıtçıbaşı and Ataca, 2005). Particularly in rural areas, becoming a wife and mother are two of the most significant aspects of women's lives (Boyacioğlu and Türkmen, 2008), and are carried out within the context of broader inter-generational kinship relations. It is difficult to assess the extent to which these traditional values are preserved among migrant communities. Although all ethnic groups in Britain are supposed to be moving (albeit at different speeds) towards lower rates of marriage and higher rates of single parenthood (Berthoud, 2000), this suggested uni-directional trend is not currently visible among Turkish migrants, who still maintain a strong ethos of procreation strictly within the boundaries of marriage. Despite the common assumption that families of immigrant origins are more traditional with respect to marriage, family formation and kinship relationships than the majority population (Shaw, 2004), there has been little qualitative research seeking to describe and understand such processes among Britain's smaller ethnic minorities.

A continuation of patrilineage, typical of Islam, and patrilocal organization of kinship have been observed among migrant Turkish communities in Germany (Nauck, 2005) and the Netherlands (van Rooij et al, 2004), and it would be fair to hypothesize similarly for London's Turks. Since most Turkish migrants come from rural settings, having never lived in an urban or 'European' context prior to their emigration, regional ties (*hemşerilik*) and family loyalties (*akrabalık*) are important to this community. Individuals tend to marry within the community, preferably with others from the same, or a similar region (Manço, 2004). Marrying-out is a rare phenomenon, and when it does occur, Turkish–British inter-ethnic marriages tend to be between individuals of relatively high social and educational status (Baltas and Steptoe, 2000). Indeed, a recent study (Enneli et al, 2005) found that young Turks and Kurds in Britain felt that it was important for them to marry within the community, and cited wider kinship concerns as the explanatory reason for this. They also accepted, and desired, general family involvement in partner choice, though they did not talk of this in terms of arranged marriages.

Among the Turkish community, there are continuing transnational links in structuring families, the most obvious of which is marriage migration. In addition, inter-generational relationships serve utilitarian functions of support and security, and transmit the culture of the society of origin (Nauck, 2005). This family-based tight-knit community both protects its members through preventing isolation, excessive marginality and juvenile delinquency (Manço, 2004) and polices and regulates the behaviour of its members. This regulation is especially important for women, who are seen as central to group identity (Mohammad,

1999) and are expected to perpetuate it through their social, procreative and nurturing acts (Akpinar, 2003).

Having children then, is often not a simple matter of individual or couple 'choice', but rather a duty and expectation, a perception strongly endorsed both by Islam (Beller-Hann, 1999; Schenker, 2000; van Rooij et al, 2004) and cultural presumptions of women's 'natural desire' to become mothers (Görgülü, 2007). The wider cultural anticipation can be exacerbated in an immigrant context, where the significance of childbearing may have particular situational consequences. As Burcu,[1] who arrived in Britain for marriage and only had a child after five years of involuntary childlessness explains in detail:

> *I remember when they were giving me a send-off* (uğurlama)*, my cousins and others, they said 'Next time we see you, it will be with your baby in your arms'. That was my dream really. To have a baby ... A baby galvanizes* (birleştirir) *families, am I explaining it well? I thought, now, here, a baby would give me and my mother-in-law* (kaynana) *a common pursuit, we would work together. And my mother would come over [from Turkey] and we would all be together, because of a baby. I was very lonely, [it was] very difficult ... And I couldn't go there [back to Turkey], because I was hesitant to turn up with no baby.* (Burcu, 27)

Kâğıtçıbaşı has written extensively about the reasons parents desire children and the value of children in Turkish society (Kâğıtçıbaşı, 1981, 1982, 1986; Kâğıtçıbaşı and Ataca 2005), and her work provides a valuable tool for interpreting such statements. She defines three categories of values that can be ascribed to children: psychological (which include the pleasure of having them around and watching them grow); economic/material/utilitarian; and social/traditional. Whilst a mixture of these categories was found to account for parents' motivations in Turkey, in general, socio-economic development and increased education was linked with a decrease in the economic and an increase in the psychological value of children. It is perhaps surprising for those from more individualistic cultures that *The Family Structure Survey 2006*, conducted by the Turkish Statistical Institute, found that the single most important child-related value, held both by men and women, was that children should look after their parents in old age. Other studies have divided up motives for childbearing as individual and social; although the former, related to the joy of childrearing (and correlating with Kâğıtçıbaşı's psychological motives), seems to be universally held in all cultures, cross-cultural differences exist in relation to the articulation of social motives (van Balen and Inhorn, 2002). Social motives, clearly present in contemporary Turkey, have also been shown to persist in Turkish migrant communities in the Netherlands (van Rooij et al, 2006), despite an absence of similar motives among Dutch respondents. From Burcu's quotation above, we can see that there is a strong emphasis on social motives, such as to galvanize the family, as well as on her personal desires and 'dream' to have a child. Such divergences regarding the role and importance of children are also, unsurprisingly, reflected in the experience of involuntary childlessness. The specific meanings of infertility and an unfulfilled desire for a child among

Turkish migrants can present particular issues for counsellors and practitioners (Gacinski et al, 2002), and elicit from the couple community-appropriate responses and reactions, which may deviate further from majority practices.

Infertility and fertility treatments

An obvious side-effect of an almost obligatory procreation mandate, and of wider family interest in reproduction, is the visibility of fertility problems. Both in Turkey, and among the diasporic Turkish community, couples are confronted by, and may themselves have, strong expectations of childbearing early in their marriages. It is likely that they will arouse implicit or explicit suspicions of infertility if these expectations are not readily met within a few years of marriage.

> *First, second, by the time it didn't work in the third month, I was already very afraid. Because I also knew that my family wanted it very much and were eagerly awaiting it, I also felt ashamed (*mahçup*) in front of them.* (Tanya, 25)

Although attention may be deflected for a time, interest in the couple's reproductive decisions are likely to persist and to increase in intensity.

> *In the beginning, to be honest, I wasn't that keen. We did not especially want children, so when people asked about it I could just respond 'İnşallah when our situation is more amenable' [then we can have children]. But it became much harder not to raise interest after a few years and everybody was saying, 'What on earth are you waiting for?'* (Hande, 30)

Once the couple (or indeed the wider family) become worried about fertility problems, they may react in a number of ways. Several interviewees mentioned that they sought prayers to help them conceive, or engaged in ceremonies to ward off the misfortune of *nazar*. Günay et al (2005), in their study of Turkish infertile couples, refer to such actions (as well as to visiting religious sites and mausoleums, consulting religious leaders and taking herbal medications) as 'traditional practices' in response to infertility, as opposed to the 'modern practices' of biomedical treatment. However, among my interviewees these practices were articulated clearly as supporting the biomedical treatments they were also seeking, rather than viewed as a replacement or alternative to them. There was an implicit understanding that such practices belong to a different paradigm and thus served different purposes to medical treatment. Allah was commonly turned to for help and guidance throughout the treatment process, and eventual treatment success (if achieved) was also appropriately celebrated by giving thanks to Allah:

> *When our son was born … two sheep were sacrificed in our village. A thousand thanks to Allah, our son was born, however many sheep are sacrificed it wouldn't be enough!* (Rizan, 37)

It deserves emphasis that religious acts or devotion among this community should not be thought of as an alternative solution, engaged in *instead of* seeking biomedical treatment. As has been found in Turkey (Albayrak and Günay, 2007), migrant Turks have high levels of consulting biomedical doctors as a response to suspected fertility problems; and there is some indication (according to members of staff at an infertility clinic in London with large numbers of Turkish patients) that they may seek medical help and present at clinics earlier than the majority population. As Schenker argues, 'When procreation fails, Islam encourages treatment ... attempts to cure infertility not only are permissible but also are a duty' (2000, p86).

Although so far there has been limited investigation of the experience of infertility or assisted reproduction among Europe's Turkish migrants (Gacinski et al, 2002; van Rooij et al, 2004, 2006, 2007; van Rooij and Korfker, Chapter 8), there is growing recognition of the importance of understanding migrants' 'local moral worlds' (to borrow a phrase from Marcia Inhorn) in order to make adequate provisions and to optimize the healthcare they are provided (van Rooij et al, 2004). For example, a questionnaire study comparing Western Turks with Turkish migrants in the Netherlands, and with Dutch men and women, found that the experience of infertility among migrants was more similar to Turkish, rather than Dutch respondents' (van Rooij et al, 2007). This was reflected especially in self-image and blame-guilt, and greater levels overall of emotional distress. It can be difficult for clinicians to appreciate, and be adequately prepared for, migrants' concerns and reactions, especially since such issues are extremely sensitive and emotive. This possible problem of a cultural mismatch was referred to in the following way by one interviewee:

> *It is very difficult to explain your problems and issues (*dert*) to the doctors here. They do not understand. We are people of different worlds.* (Tanya, 25)

Moreover, two interviewees related negative experiences in their pursuit for assisted reproduction treatment in London, which were directly linked, in their perceptions, with their immigrant status:

> *I was not offered a test tube baby here. These doctors are not interested in helping us, because they would rather that foreigners don't have any more children in their country. So I had to go to Turkey for my treatment, which I am very pleased about, my doctors there were very humane (*insancıl*) and friendly (*cana yakın*) and looked after me very carefully.* (Kiraz, 25)

We can see that Kiraz has interpreted what was probably a straightforward medical decision as a political one, perhaps due to her own sensitivities. Although Dilek did not see the interaction in such politicized terms, she found the doctor's insistence for her to lose weight in order to maximize the chances of spontaneous conception (which is not an uncommon requirement in cases of unexplained infertility where women are over a certain body mass index) as an inappropriate expectation, born from the doctor's lack of knowledge about her culture's norms:

> *They kept me on a waiting list for ages, and every time I went to the doctor,*
> *she kept saying you have to lose more weight, lose more weight [before I can*
> *accept you for treatment] … [Laughs] My dear, my mother is like this, my*
> *father is like this, my entire family is [big] like this: so this is not the reason*
> *for my problems! We [Turks] are made meaty (*etli butlu*). (Dilek, 33)*

Both of these cases, although in very different ways, highlight potential mis-understandings, or miscommunications, which can arise when doctors and patients are acting within different cultural frameworks. Regardless of the objec-tive facts of each situation, it is unfortunate that the women constructed and maintained such difference-based accounts in order to explain what they perceived as their unsatisfactory interactions with the British healthcare system.

In fact, most of my respondents mentioned that where possible, they would prefer to see a Turkish doctor, and several had indeed found Turkish doctors, either here or in Turkey, with whom they had established a good rapport.

> *I went to a Turkish doctor [in London], and I explained to him that my*
> *psychology has been ruined (*psikolojim bozuldu*) because for months*
> *nothing had happened. I felt terrible about myself. He calmed me down and*
> *explained that there are solutions. (Tanya, 25)*

Hande, who is currently still childless after two cycles of IVF treatment in London, had decided that her chances of success might be improved if she went for treatment to Turkey:

> *I do have a strong belief that it is going to work for us sometime in the*
> *future. We are planning to go to Turkey for treatment next year, and we*
> *have heard very good things. We know many people who were told they*
> *couldn't have children here, and only had their children when they went to*
> *Turkey. (Hande, 30)*

These women relied on female friends and relatives (both in London and in Turkey), as well as the Turkish media, which they all followed, to be sources of accurate information regarding infertility and assisted reproduction. It is there-fore possible that their opinions about fertility treatments were somewhat shaped by the particular discourses employed within the Turkish media. Some of these discourses draw on the metaphor of 'infertility as a war to be fought against' where assisted reproduction technologies are modern weapons, and overwhelm-ingly present successful endings as the standard treatment narrative, albeit emphasizing the importance of patience and endurance in order to win such a war (Görgülü, 2007). IVF is presented as a wondrous, yet routinely exercised, medical solution to a medical problem, and great national pride is taken in Turkey's 'test-tube baby stars' and their state-of-the-art clinics. Though it is not possible to draw any causal inferences from such a limited study, it was never-theless interesting that some interviewees articulated views about IVF, and specifically about IVF treatment in Turkey, which echoed these media represen-tations.

Typically, as is also the case in Turkey (Günay et al, 2005), my interviewees reported that it was themselves, rather than their husbands, who initiated medical treatment and underwent initial medical tests. Throughout the assisted reproduction processes (partly as imposed by the nature of such treatments), women maintained a greater level of interest and involvement. Although all of the women I interviewed referred to infertility treatment very much as something a couple had to undergo together, offering mutual support and understanding along the way, they at times had to negotiate the practicalities of regular hospital appointments by themselves or with female relatives, who would act as substitute supporters.

> *Because of my husband's work situation, he could not always come to the hospital with me. My sister would sometimes come with me, which is nice... When I had an appointment [which my husband couldn't be at] and came out, I would have ten missed calls from him, he would be so worried.* (Dilek, 33)

These women reported that they had a great influence on treatment decisions, and that they would often become informed themselves before guiding their husbands through the possible options. Several of my respondents echoed findings that it is women who assume the role of risk assessors and arbiters regarding which risks are worth taking in the course of fertility treatment (Becker and Nachtigall, 1994).

> *After our first try ... he saw that I was exhausted (bitik) and my husband did not want to continue. He was very sad to see me in such a state, he did not think I could handle such a big disappointment again, he felt our chances were low and he did not want to put me through that... But I made the decision that we should keep trying, and when I told him so he was willing to go along with me.* (Burcu, 27)

Generally, close family members and friends were informed that treatment was being sought; in some cases this disclosure was unavoidable since the wider family were required to provide financial help, or to facilitate various other practicalities (such as time off work) necessitated by the treatment. In other cases, family members had suspected fertility problems and had themselves initiated communication about it. Repeated communication and exposure were felt to normalize these subjects.

> *There isn't any shame in talking about this treatment I think. Perhaps some old fashioned (geri kafalı) people might feel like that, but to me it is very natural (doğal) and that is the way my family perceive it, they too are used to it.* (Dilek, 33)

> *In the beginning I was a little embarrassed to talk about it. Now I can talk quite comfortably, and as I talk I feel more comfortable.* (Meryem, 30)

It is quite striking that Meryem should say that she talks 'quite comfortably', considering that she later informed me that she had not disclosed her husband's azoospermic diagnosis to anyone. For her, as for some other interviewees, there were strong lines of demarcation regarding which aspects of infertility and fertility treatments were appropriate areas for discussion and which were private matters for the husband and wife. Certain particular details, such as whether it was male or female factor infertility, were most likely to have been hidden from or actively misrepresented to others.

> *I talk to my sister, and my mother a lot about this, they are very understanding. Some people will ask whether it is because of me or my husband, but I find that very hurtful, it is a very private matter to ask about. I just say it must be something about both of us.* (Rizan 37)

> *I said we are going to the doctor's, and we will see if they can help us. I pretended to be more naive because I did not want to go into details that might embarrass me or my husband.* (Fatma, 29)

Some of the women in my study reported that their husbands found the treatment process more difficult, and related this to the greater stigmatization of male infertility, as well as to men's reduced ability and opportunity to communicate about treatment. This latter point is succinctly encapsulated by Kiraz:

> *It is much more difficult for men. They are closed-up.* (Kiraz, 25)

Indeed, one can argue that cultural interpretations in which both virility and fertility are directly connected to manhood (Inhorn, 2004), as is the case in Turkey as well as more broadly in the Middle East, mean that fertility problems are experienced as emasculating and highly stigmatized conditions for men. Moreover, migrant Turkish men seem to show even greater levels of emotional distress than men in Turkey, as was found by van Rooij et al (2007) in their previously mentioned comparative study of infertile Turkish migrants in the Netherlands.

> *My husband entered a deep depression when he was told that he was infertile. He could not cope with this at all, he became very withdrawn.* (Oya, 34)

In addition, low levels of awareness regarding male infertility in parts of Turkish society (Üner, 2004), the corresponding attribution of infertility largely upon women (Günay et al, 2005), and a commonly mistaken equation of male virility with fertility, as has been explored in detail in Delaney's 'seed and soil' thesis (1987, 1991), may exacerbate men's sense of profound shock and shame in the event of such a diagnosis. Yüksel et al (1996; cited in van Rooij et al, 2004) found that when Turkish men in Germany were told about their infertility, almost half of them denied or did not accept this. Such findings not only illustrate the need for cultural sensitivity in clinical practices, but also highlight that the experience of infertility and fertility treatments may be influenced by the presence of alternative procreation beliefs among migrant communities. In order to unravel the various threads of cultural complexity regarding gendered interpretations and experiences, it is useful to refer to a rich body of ethnographic literature about infertility and reproductive technologies in Muslim Middle Eastern countries (Inhorn, 1996, 2003a, 2003b, 2003c, 2004, 2006a, 2006b, 2006c, 2007; Clarke, 2006a, 2006b). Although it would be culturally short-sighted to assume that views within London's Turkish community exactly replicate those in the Middle East, simply as a consequence of common Islamic religious beliefs, those studies nevertheless provide invaluable explorations of some of the commonly observed behaviours among such migrant communities.

For example, it has been observed by Inhorn (2003c) that women typically bear the social burden of infertility, even when their husbands are the ones with physiological factors. This is precisely what Oya was prepared to do if it was going to improve her and her husband's situation.

> *I comforted him a little by telling him that no-one needs to know about this, it can be just between us, it has nothing to do with anyone else. I was quite prepared to say it was something to do with me. I wanted him to stop mourning, because I hate to see him so sad, and also because I wanted us to start thinking productively about what there is that we could do.* (Oya, 34)

Lober asks why a woman would agree to undergo IVF (which is physically invasive and difficult for her) in circumstances of male infertility. She has called this a 'patriarchal bargain', and argues that a woman may thus attempt to '[resolve] a situation in which she has limited options in the best way she can' (1989, p30). Perhaps a similar logic could be extended to why a woman would assume her husband's infertility. Indeed, it is difficult to deny the palpable presence among this community of a cultural pressure that exerts the importance of biological reproduction and descent, especially for men. Although it would be too simplistic to attribute Oya's behaviour solely to patriarchal pressures and to ignore her agency and expressed motivation of wanting to help her husband cope, there is certainly an element of bargaining to her actions. It was only through her concerted efforts to enable her husband to deal with his infertility (albeit by covering it up entirely) that Oya could persuade him to think about resolving their problem of childlessness; in this instance her overt assumption of responsibility for the infertility was translated into responsibility and decisive power regarding treatment options, as I will return to later.

As Inhorn (2007) warns, however, we should not ignore that men too suffer 'both somatically and psychically' in the treatment process, not only because of the stigma associated with male infertility but also as a result of what the treatment requires of them; producing a sperm sample (by masturbation), reported as a time of heightened anxiety by men undergoing IVF, can be especially difficult for Muslim men because of cultural connotations and prohibitions. It was therefore shocking to me that when precisely such a question was raised at a British Infertility Counselling Association (BICA) Study Day in London, in 2006 (regarding in that instance a particular Pakistani Muslim man's inability to produce a sperm sample), the counsellor leading the session remarked that, under these circumstances, one could only begin to doubt the authenticity of this man's desire to have a child. Undoubtedly, greater knowledge of ethnic minorities' particular cultural concerns will enable better understandings of their reactions and attitudes towards specific aspects of fertility treatment. This is seen to be a particularly thorny issue with regard to the use of donor gametes.

Although some Shiite opinions are permissive of the use of donor gametes within fertility treatments (as practised in Iran and Lebanon), there is broad consensus among the Sunnis (the branch of Islam to which a very large majority of Turks belong) that 'medical interventions in human reproduction should restrict themselves to a husband and (one) wife couple' (Clarke, 2006b, p26). Saim Yeprem, from Turkey's Higher Council of Religious Affairs, explains the grounds of this religious prohibition: 'for a child to be a legitimate one, Islam asserts, they should belong to the legitimate spouse' (2007, p44). Moreover, since donor gametes contradict the Islamic principles of the preservation of genealogy, they 'cause confusion regarding the authentication of the child's generation' (2007, p46). 'Preserving the "origins" of each child – meaning its relationship to a known biological mother and father – is considered not only an ideal in Islam, but a moral imperative' (Inhorn, 2006c, p440). Although Turkey is a secular country, its moral and ethical decisions are largely informed and moderated by Islamic morality. In this instance, resulting in clear restrictive guidelines, according to which all forms of third party involvement (donor sperm, donor eggs, surrogacy) are illegal. In fact, a legal document defines assisted reproduction treatments as:

> *Procedures which involve assisting the fertilization of the prospective* mother's egg *with* her husband's sperm *in various ways, enabling them to fertilize outside of the body when necessary, and transferring the gametes or the embryo back to the prospective mother's genital organs, and are procedures accepted as treatment methods by modern medicine.* (Ministry of Health of Turkey, my emphasis)

Such a definition clearly places the marital unit as practically indispensable and morally central to infertility treatments; it also leaves single women and homosexual couples, as well as those in traditional marriages with the misfortune to require donor gametes, firmly outside the technologies' reach.

However, this is not to suggest that no Turkish couples would consider using donor gametes. Clearly individuals do not always abide by religious teachings, cultural norms or state regulations. Although there are no official figures, newspapers in Turkey sporadically run stories of Turkish infertile couples who visit neighbouring Cyprus and Greece (where donor sperm treatments are readily available), only to come back 'importing' Greek or Danish sperm, accusing them of moral degeneracy and national threat. Indeed, a recent study conducted on infertile men and women in Turkey showed that 23 per cent would accept donor eggs; 15.1 per cent would accept surrogacy; and 3.4 per cent would accept donor sperm if it was available and medically indicated (Baykal et al, 2008). Among the general Turkish population, there was staggeringly high acceptance of donor eggs as a medical treatment, with over half of the women and nearly two-thirds of the men replying that they thought their religion would allow it (Isikoglu et al, 2006). These studies illustrate the differences in attitudes towards eggs and sperm, yet they also demonstrate that even with strict cultural codes of practice, there is room for individual moral manoeuvring.

In Oya's case, she and her husband chose to resolve their infertility by using donor sperm. I am told (by clinicians in London) that this is an unusual treatment for Turkish migrants to pursue, and they are indeed extremely secretive about it. Their secrecy of donor conception is necessitated, since they are also secretive about male factor infertility (as expressed above), yet she explains their decision for secrecy as a way to protect their child within the community, which they feel is not yet ready to embrace donor-children:

> We love our son, and we are very happy to have him exactly as he is. I am not ashamed of him at all, but neither would I want to make him live through undue difficulties. It is undeniable that there would be talk about him if we told people, they may even call him nasty names. For me, this is completely unacceptable... It would absolutely break my heart. (Oya, 34)

However, the fact that they have a donor-sperm conceived child is significant, and should serve as a testament that migrant couples may go against the cultural norms and prohibitions of their community, and decide to make use of available technologies in order to have the child they so desire.

Conclusion

Here I have drawn a portrait of London's Turkish community, providing snap-shots from their daily life and supplying details of their views on families and the value of children, in order to create a framework within which understandings of their attitudes about and experiences of involuntary childlessness may be developed. I have argued for caution against blunt categorizations of 'ethnic minorities', and against assumptions of cultural homogeneity between migrant populations and their home contexts, as well as within migrant populations which may be constituted of different sub-groups of people. Since my interview sample was small, I do not want to make broad generalizations; however, I have used the data from those interviews to draw out and highlight some pertinent aspects of Turkish women's experiences of infertility and fertility treatments in London.

The Turkish community in London have strong family oriented values, and maintain close links, both locally and transnationally, with inter-generational kin. Children are desired both because of the individual psychological pleasures they are seen to bring to their parents, and because of their traditional and social values, such as creating family cohesion and continuity. In such a context, there may be great interest from wider kin regarding a couple's reproductive decisions, as well as pressures and expectations on them to have children early in their marriage. One obvious consequence of this focused gaze is that childless couples arouse (warranted or perhaps even premature) suspicions of infertility from their social surroundings. A common reaction to such worries is the seeking of biomedical treatment. Women tend to lead the way in this regard and also to assume a greater responsibility for treatment decisions. Gendered sensitivities and the continued (if diminished) existence of alternative procreation models often mean that male infertility is a difficult diagnosis to acknowledge and accept. Although there are strong religious and cultural prohibitions against the use of donor gametes, this does not translate to a blanket refusal of such medical possibilities.

Although assisted reproduction technologies may pose a range of difficulties (financial, practical, emotional, psychological, as well as moral), for those undergoing them, to paraphrase Simpson (2004, p230), they are ultimately embraced because they address the fundamental desires of men and women to become parents. Whether it makes a difference if such desires are internalizations of cultural pressures and expectations is a question that has little bearing on the perceived pain and immediacy of involuntary childlessness:

> *I would not wish having to go through all of this [fertility treatments] even on my enemies. But … you just get on with it. At the end, a woman would do whatever is within her means, anything to become a mother, wouldn't she?* (Burcu, 27)

Acknowledgements

This research was undertaken as part of an ESRC-funded PhD project, based at the Centre for Family Research, Faculty of Social and Political Sciences, University of Cambridge. I would like to thank Alex Broadbent, Charlotte Faircloth, Susan Golombok, Martin Richards and the editors for their useful comments on earlier versions of this paper. I am grateful to all the women who generously gave their time and stories for this research.

Notes

1   All names, and some identifying information, have been changed in order to protect privacy and confidentiality.

References

Akpinar, A. (2003) 'The honor/shame complex revisited: Violence against women in the migration context', *Women's Studies International Forum*, vol 26, pp425–442
Albayrak, E. and Günay, O. (2007) 'State and trait anxiety levels of childless women in Kayseri, Turkey', *The European Journal of Contraception and Reproductive Health Care*, vol 12, pp385–390
Balen, F. van and Bos, H. M. W. (2004) 'Infertility, culture, and psychology in worldwide

perspective', *Journal of Reproductive and Infant Psychology*, vol 22, pp245–247

Balen, F. van and M. C. Inhorn, (2002) 'Interpreting infertility: A view from the social sciences', in M. C. Inhorn, and van Balen, F. (eds) *Interpreting Infertility: Childlessness, Gender and Reproductive Technologies in Global Perspective*, UCLA Press, Berkeley, CA

Baltas, Z. and Steptoe, A. (2000) 'Migration, culture conflict and psychological well-being among Turkish–British married couples', *Ethnicity & Health*, vol 5, pp173–180

Baykal, B., Korkmaz, C., Ceyhan, S. T., Goktolga, U. and Baser, I. (2008) 'Opinions of infertile Turkish women on gamete donation and gestational surrogacy', *Fertility and Sterility*, vol 89, pp817–822

Becker, G. and Nachtigall, R. D. (1994) '"Born to be a mother": The cultural construction of risk in infertility treatment in the US', *Social Science & Medicine*, vol 39, pp507–518

Beller-Hann, I. (1999) 'Women, work and procreation beliefs in two Muslim communities', in P. Loizos and P. Heady (eds) *Conceiving Persons: Ethnographies of Procreation, Fertility and Growth*, Atholone Press, London

Berry, J. (1997) 'Immigration, acculturation, and adaptation', *Applied Psychology*, vol 46, pp5–34

Berthoud, R. (2000) *Family Formation in Multi-Cultural Britain: Three Patterns of Diversity*, Institute for Social and Economic Research, University of Essex

Boyacioğlu, A. Ö. and Türkmen, A. (2008) 'Social and cultural dimensions of pregnancy and childbirth in eastern Turkey', *Culture, Health & Sexuality*, vol 10, pp277–285

Clarke, M. (2006a) 'Islam, kinship and new reproductive technology', *Anthropology Today*, vol 22, pp17–22

Clarke, M. (2006b) 'Shiite perspectives on kinship and new reproductive technology', *Review of the International Institute for the Study of Islam in the Modern World*, vol 17, pp26–27

Delaney, C. (1987) 'Fields of honor, seeds of shame', in D. D. Gilmore (ed) *Honor and Shame and the Unity of the Mediterranean* (A Special Publication of the American Anthropological Association, no 22), American Anthropological Association, Washington, DC

Delaney, C. (1991) *The Seed and the Soil: Gender and Cosmology in Turkish Village Society*, University of California Press, Berkeley, CA

Enneli, P., Madood, T. and Bradley, H. (2005) 'Young Turks and Kurds: A set of "invisible" disadvantaged groups', Report supported by the Joseph Rowntree Foundation, University of Birmingham

Gacinski, L., Yüksel, E. and Kentenich, H. (2002) 'The unfulfilled desire for a child of oriental couples: Infertility counselling and treatment of Turkish immigrants in Germany', in B. Strauss (ed) *Involuntary Childlessness: Psychological Assessment, Counselling and Psychotherapy*, Hogrefe & Huber Publishers, Seattle, WA

Görgülü, B. (2007) 'Interpreting IVF from a Foucauldian perspective', Graduate School of Arts and Social Sciences, Sabanci University, Istanbul, Turkey

Günay, O., Çetinkaya, F., Naçar, M. and Aydin, T. (2005) 'Modern and traditional practices of Turkish infertile couples', *The European Journal of Contraception and Reproductive Health Care*, vol 10, pp105–110

Inhorn, M. C. (1996) *Infertility and Patriarchy: The Cultural Politics of Gender and Family Life in Egypt*, University of Pennsylvania Press, Philadelphia, PA

Inhorn, M. C. (2003a) 'Global infertility and the globalization of new reproductive technologies: Illustrations from Egypt', *Social Science and Medicine*, vol 56, pp1837–1851

Inhorn, M. C. (2003b) *Local Babies, Global Science: Gender, Religion, and In Vitro Fertilization in Egypt*, Routledge, New York

Inhorn, M. C. (2003c) '"The worms are weak": Male infertility and patriarchal paradoxes in Egypt', *Men and Masculinities*, vol 5, pp236–256

Inhorn, M. C. (2004) 'Middle Eastern masculinities in the age of new reproductive technologies: Male infertility and stigma in Egypt and Lebanon', *Medical Anthropology Quarterly*, vol 18, pp162–182

Inhorn, M. C. (2006a) '*Fatwas* and ARTs: IVF and gamete donation in Sunni v. Shi'a Islam', *The Journal of Gender, Race and Justice*, vol 9, pp291–317

Inhorn, M. C. (2006b) '"He won't be my son": Middle Eastern Muslim men's discourses of adoption and gamete donation', *Medical Anthropology Quarterly*, vol 20, pp94–120

Inhorn, M. C. (2006c) 'Making Muslim babies: IVF and gamete donation in Sunni versus Shi'a Islam', *Culture, Medicine and Psychiatry*, vol 30, pp427–450

Inhorn, M. C. (2007) 'Masturbation, semen collection and men's IVF experiences: Anxiety in the Muslim world', *Body & Society*, vol 13, no 3, pp37–53

Isikoglu, M., Senol, Y., Berkkanoglu, M., Ozgur, K., Donmez, L. and Stones-Abbasi, A. (2006) 'Public opinion regarding oocyte donation in Turkey: First data from a secular population among the Islamic world', *Human Reproduction*, vol 21, pp318–323

Kâğıtçıbaşı, Ç. (1981) 'Value of children, women's role and fertility in Turkey', in Abandan-Unat, N. (ed) *Women in Turkish Society*, E. J. Brill, Leiden

Kâğıtçıbaşı, Ç. (1982) *The Changing Value of Children in Turkey*, East–West Population Institute, Honolulu

Kâğıtçıbaşı, Ç. (1986) 'Status of women in Turkey: Cross-cultural perspectives', *International Journal of Middle East Studies*, vol 18, pp485–499

Kâğıtçıbaşı, Ç. (2005) 'Autonomy and relatedness in cultural context', *Journal of Cross-Cultural Psychology*, vol 36, pp403–422

Kâğıtçıbaşı, Ç. and Ataca, B. (2005) 'Value of children and family change: A three-decade portrait from Turkey', *Applied Psychology*, vol 54, pp317–337

Lober, J. (1989) 'Choice, gift or patriarchal bargain? Women's consent to in vitro fertilization in male infertility', *Hypatia*, vol 4, pp23–36

Manço, U. (2004) *Turks in Europe: From a Garbled Image to the Complexity of Migrant Social Reality*, Centrum Voor Islam in Europa (C. I. E.), available online at: http://www.flwi.ugent.be/cie/umanco/umanco5.htm

Mohammad, R. (1999) 'Marginalisation, Islamism and the production of the "Other's" Other', *Gender, Place and Culture*, vol 6, pp221–240

Nauck, B. (2005) 'Intergenerational relations in Turkish families in Germany', in J. Pflegerl and S. Trnka (eds) *Migration and the Family in the European Union*, Austrian Institute for Family Studies, Vienna

Nazroo, J. Y. (ed) (2006) *Health and Social Research in Multiethnic Societies*, Routledge, London

Rooij, F. B. van, van Balen, F. and Hermanns, J. M. A. (2004) 'A review of Islamic Middle Eastern migrants: Traditional and religious cultural beliefs about procreation in the context of infertility treatment', *Journal of Reproductive and Infant Psychology*, vol 22, pp321–331

Rooij, F. B. van, van Balen, F. and Hermanns, J. M. A. (2006) 'Migrants and the meaning of parenthood: Involuntary childless Turkish migrants in The Netherlands', *Human Reproduction*, vol 21, pp1832–1838

Rooij, F. B. van, Balen, F. van and Hermanns, J. M. A. (2007) 'Emotional distress and infertility: Turkish migrant couples compared to Dutch couples and couples in Western Turkey', *Journal of Psychosomatic Obstetrics and Gynaecology*, vol 28, pp87–95

Schenker, J. G. (2000) 'Women's reproductive health: Monotheistic religious perspectives', *International Journal of Gynaecology & Obstetrics*, vol 70, pp77–89

Shaw, A. (2004) 'Immigrant families in the UK', in J. Scott, J. Treas and M. Richards (eds) *The Blackwell Companion to the Sociology of Families*, Blackwell Publishing, Oxford

Simpson, B. (2004) 'Acting ethically, responding culturally: Framing the new reproductive and genetic technologies in Sri Lanka', *The Asia Pacific Journal of Anthropology*, vol 5, pp227–243

Üner, Ö. (2004) 'A study on social and affective dimensions of assisted reproduction technology (ART) by women in Turkey', *The Graduate School of Social Science*, Middle East Technical University, Ankara, Turkey

Yeprem, S. (2007) 'Current assisted reproduction treatment practices from an Islamic perspective', *Reproductive BioMedicine Online*, vol 14, pp44–47

Yüksel, E., Siemann, A. and Kentenich, H. (1996) 'Kinderlosigkeit Türkischer Paare in der Migration: Ein multifakorielles problem' [Childlessness among migrant Turkish couples: A multi-factoral problem], *Geburtshilfe und Frauenheilkunde*, vol 56, pp188–190

Web references

Ministry of Health of Turkey, Turkish Assisted Reproduction Treatment Centres Directorate, Official Gazette, 31 March 2001, Number 24359, available from: www.istanbulsaglik.gov.tr/w/mev/mev_yon/uremeye_yardimci.pdf

Turkish Embassy 2008, 'Turks in London', available from: http://turkishembassylondon.org/canon/turks.htm

Turkish Statistical Institute (TÜRKSAT), The Family Structure Survey 2006, available from: www.turkstat.gov.tr/VeriBilgi.do?tb_id=64&ust_id=11

UK Census 2001, available from: www.statistics.gov.uk/census2001/census2001.asp

UK Census 1991, available from: www.lib.gla.ac.uk/Resources/Databases/1991.shtml

8

Infertile Turkish and Moroccan Minority Groups in the Netherlands: Patients' Views on Problems within Infertility Care

Floor van Rooij and Dineke Korfker

Introduction

Studies conducted in Western and non-Western societies[1] indicate that in many contexts, involuntary childlessness is a sensitive topic and a source of personal concern and social stress (Inhorn and van Balen, 2002). However, people (particularly women) living in non-Western societies appear to experience more severe social consequences of involuntary childlessness, such as social stigmatization and exclusion, marital or social violence and more emotional distress, than people in Western societies (Daar and Merali, 2001; van Balen and Inhorn, 2002). These observed differences in social consequences and emotional distress are frequently related to a stronger pronatalist culture (i.e. having children is the undisputed social norm) in non-Western than in Western societies (van Balen and Inhorn, 2002; Dyer, 2007) and to differences in cultural beliefs about procreation and the causes of infertility (see also van Balen, Chapter 2). Such cultural differences might also influence the help-seeking behaviour of infertile couples (Gerrits, 1997; Guntupalli, 2004; Nahar, 2007; van Rooij, 2008).

Little is known about the help-seeking behaviour of infertile non-Western couples in the Netherlands. No information is available regarding the number of non-Western couples living in the Netherlands using infertility services, as data about the ethnic background of patients are not routinely collected in the Netherlands (Evenblij et al, 2004). Furthermore, little is known about the experiences of accessing healthcare among infertile non-Western men and women living in the Netherlands.[2] In a small-scale study of migrant Ghanaian women living in Amsterdam, Yebei (2000) describes the way in which infertility leads to stigmatization, divorce, abuse, resentment and loss of social status and self-esteem. This study also highlighted the difficulties that women encountered in accessing fertility treatment in the Netherlands, as many of them were 'undocumented' migrants and economically disadvantaged. Dissatisfaction with fertility treatment received was also evident, arising from a language barrier between women and doctors, lack of information about their treatment and unsympathetic attitudes.

Prior to the studies reported in this chapter, the experiences of Turkish and Moroccan migrants in the Netherlands experiencing fertility problems was largely unexplored, although some work on Turkish migrants in Germany has shown that Turkish migrants are more willing to start fertility treatment at a younger age, report a stronger desire to have a child, set fewer time limits for in vitro fertilization (IVF) treatment, have a lower willingness to adopt, and report higher levels of emotional distress as compared to infertile indigenous Germans (Yüksel, 1995; Kentenich and Yüksel, 1997; Gacinski et al, 2002). Studies of minority ethnic groups in other developed countries also show several possible problems in accessing and receiving appropriate infertility treatment, relating to a lack of communication support and access to economic resources (e.g. Culley et al, 2006 (UK); Inhorn and Fakih, 2006 (US)). In countries that lack a national health insurance system, an important prerequisite to accessing healthcare is affordability. As non-Western migrants are in general overrepresented in the lower socio-economical strata (Inhorn and Fakih, 2006; Nazroo, 2003; Statistics Netherlands, 2008), affording infertility treatment might be a particular difficulty for these groups (see also Inhorn and Fakih, 2006; Inhorn et al, Chapter 11). In countries like the Netherlands, however, a substantial part of the available health treatments are covered by compulsory health insurances. Therefore, in general, all insured people should be able to have most treatments. With respect to fertility treatments, this means that all insured people have access to examinations and to most treatments up to a certain number (most insurances cover three cycles of IVF treatment for women up to the age of 40–45; age limits vary between insurances and hospitals (www.freya.nl)).[3] However, as Yebei (2000) points out, migrants who are 'undocumented' will face considerable financial barriers to treatment in the Netherlands as they are unable to use the national health insurance cover and are unlikely to be able to afford direct payments for biomedical treatments. According to Yebei, undocumented Ghanaian women opted for alternative infertility treatments such as herbs and spiritual healing.

Studies about the healthcare utilization of migrants in the Netherlands in general show that, even when adjusted for age, gender, education, insurance status, health status, and level of urbanization, the utilization of services differs across ethnic groups. People from most major migrant groups make more use of general practitioners (GPs) (gatekeepers to specialized care) than indigenous Dutch people (Uiters et al, 2006). Interestingly, Moroccans show a pattern of high GP consultations and a relatively low utilization of other healthcare services, indicating potential problems with accessing specialized care (Uiters et al, 2006). Uiters et al (2006) suggest that this might be a consequence of a poorer quality of consultations, with patients having to make several visits to the GP in order to obtain a referral.

This poorer quality of consultations is possibly a consequence of communication problems between patients and caregivers. Studies of access to infertility care by Yebei (2000), Culley et al (2006) and Inhorn and Fakih (2006) found that infertile migrant people who are less fluent in the language of the host country have more difficulties with accessing fertility services and related relevant information than those who do know the dominant language. Besides these common communication problems, differences in cultural understandings of healthcare appear to play a role (e.g. Yebei, 2000; Culley et al, 2006). There may be, for example, differences in explanatory models of health and illness and cultural differences in patients' preferences for doctor–patient relationships (Schouten and Meeuwesen, 2006; Korfker et al, 2008). Such problems might not only lead to poorer quality of care and less access to healthcare and relevant information, but also to inappropriate use of healthcare services, non-compliance, lower levels of satisfaction among minority patients, and feelings of discrimination in service provision (e.g. Ferguson and Candib, 2002; Culley et al, 2006; Meeuwesen et al, 2006; Uiters, 2007; Harmsen et al, 2008).

In this chapter we will focus on the problems facing infertility patients of Turkish and Moroccan descent in the Netherlands. We discuss the findings of two research projects, which explored the needs of Turkish and Moroccan migrant communities (van Rooij et al, 2006, 2007; Korfker et al, 2008; van Rooij, 2008). We begin by briefly describing these two communities and the social meanings of infertility among their members. We then outline the methodology of the two studies and go on to discuss some key themes which emerged from our research.

Turkish and Moroccan migrants in the Netherlands

Turkish and Moroccan minority groups are two of the largest ethnic minority groups in the Netherlands (368,600 and 329,493 respectively) (Statistics Netherlands, 2007). Both groups share a common migration history in that in the 1960s the first Turkish and Moroccan people were invited by Dutch companies to come to the Netherlands because of a shortage in the labour market, particularly in the lower segment (SCP/WODC/CBS, 2005). Most Turkish labourers were recruited in Central Anatolia, a region in Turkey, with half coming from villages, 5 per cent from large cities, and the remainder from smaller cities (Kaya, 2006). A significant proportion of the Moroccan migrants also came from rural areas – predominantly the Riff Mountains. A majority (85 per cent) of Moroccan migrants have a Berber background, while others have an Arabic background (van Heelsum, 2003). The influx of Turkish and Moroccan migrants to the Netherlands increased rapidly during the 1960s. When labour recruitment was brought to an end in 1974, the flow of new Turkish and Moroccan migrants into the Netherlands continued due to family reunification, marriage, asylum requests and informal channels (de Valk, 2001; Kaya, 2006; Statistics Netherlands, 2008). Over recent years the immigration of Turkish and Moroccan people to the Netherlands has been mainly the consequence of the high numbers of Turkish and Moroccan migrants (first and second generation) finding their spouses in their country of origin (Hooghiemstra, 2003; SCP/WODC/CBS, 2005). Since 2003, however, the number of Turkish and Moroccan people coming to the Netherlands on marriage has decreased (van Huis, 2008).

Turkish and Moroccan people in the Netherlands have a relatively low level of education in comparison with Dutch people, and women have lower educational levels than men. Nevertheless, the second generation is catching up when compared to Dutch people and the differences between men and women are becoming less pronounced (SCP/WODC/CBS, 2005; Kaya, 2006). Furthermore, about three-quarters of all Turkish migrants and about half of all Moroccan migrants report that they sometimes have difficulties with having a conversation in Dutch. Identification with their own ethnic groups is high: almost half of the Turkish migrants and 40 per cent of the Moroccan migrants identify themselves predominantly as Turkish or Moroccan, while only 12 per cent of the Turkish migrants and 17 per cent of the Moroccan migrants feel themselves predominantly Dutch. Two-thirds of the Turkish migrants and just over half of the Moroccan migrants have more contact with people with the same ethnic background than with the indigenous Dutch populations (SCP/WODC/CBS, 2005).

A further similarity between the two groups is in religious background. Sunni Islam is the religion of the vast majority of both Moroccan and Turkish minorities. However, Moroccan people in the Netherlands appear to be more 'orthodox' in their religious beliefs than Turkish people in the Netherlands (SCP/WODC/CBS, 2005). Within Sunni Islam, procreation is strongly encouraged and is seen as one of the five basic goals of Islamic law. Having or not having children is believed to be God's Will. Nevertheless, infertile people are also encouraged by Islam to seek medical help (Husain, 2000). The exception to this is that treatment involving third party gametes is not allowed according to religious regulations within Sunni Islam (e.g. Inhorn, 1996, 2003; Meirow and Schenker, 1997; Serour, 1998, 2001). It might well be that this is more important for Moroccans than for Turkish migrants, as they attach more value to following religious rules.

Little is known about how infertility is understood in Moroccan culture, both within Morocco itself or among migrant communities. A study in Belgium exploring contraceptive use among Moroccan women, however, mentions that having children is a strong social norm; that involuntary childlessness is a very heavy burden which leads to stress and is a very sensitive subject (Lodewijckx and Hendrickx, 1996). There is some evidence that this is also applicable to infertile Moroccan minorities in the Netherlands (Korfker et al, 2008).

With respect to Turkish migrants in the Netherlands, previous publications by van Rooij et al (2006, 2007) and van Rooij (2008) show that infertility has a high impact on the lives of most, but not all, involuntarily childless Turkish men and women and that levels of emotional distress are higher among infertile Turkish minorities than they are among indigenous Dutch infertility patients. This might be due to the stronger pronatalist norm among Turkish minorities and the higher social importance of having children than among indigenous Dutch people (van Rooij et al, 2006). These findings are similar to those found in studies of Turkish minorities in Germany (Yüksel, 1995; Kentenich and Yüksel, 1997) and the UK (see Gürtin-Broadbent, Chapter 7).

Methods and study populations

The findings from two distinct projects form the basis of this chapter (van Rooij et al, 2006, 2007; Korfker et al, 2008; van Rooij, 2008). The project by Korfker and colleagues (2008) (hereafter referred to as 'Study One'), focused on the problems Turkish and Moroccan infertility patients perceived within the Dutch healthcare system. In 'Study Two', van Rooij and colleagues investigated the parenthood motivations, experiences, psychosocial consequences and help-seeking behaviour of involuntary childless Turkish migrants in the Netherlands (van Rooij et al, 2006, 2007; van Rooij, 2008). This study also explored the experiences of involuntarily childless Turkish people within the Dutch healthcare system. All participants from both studies had been or were visiting infertility services in the Netherlands because of their fertility problems.

Study One included an interview study with 21 Turkish and 21 Moroccan women, and 8 Turkish and 31 Moroccan men. Men and women were recruited via fertility doctors and from community settings. They were interviewed separately (using a semi-structured interview schedule) by migrant health workers, matched to the participants by gender and ethnicity. Study Two included interviews with 11 couples and nine women from Turkish migrant communities. The interviews were carried out by a bilingual indigenous Dutch female researcher. All but two interviews were audiotaped and transcribed verbatim (see van Rooij, 2008). The participants in Study Two were recruited both within and outside the biomedical healthcare system (see also, van Rooij et al, 2006; van Rooij, 2008).

The Turkish migrant men and women in Study One had less formal education than the Turkish participants in the second study. Both studies included participants who were born in the Netherlands, participants who had been resident in the Netherlands for more than ten years and participants who had migrated within the previous ten years. The vast majority of the participants in both samples belonged to the Sunni Islam faith. In both studies people with female factor infertility, male factor infertility, female and male factors combined and 'unexplained infertility' were represented. However, not all Turkish and Moroccan migrant participants in Study One were able to tell the researchers the reason for their infertility as they hadn't understood this. Some of the key themes from these studies are presented below, with an emphasis on the experiences of our participants in the Dutch healthcare system.

Experiences with healthcare services

Our discussion of the encounter of Turkish and Moroccan infertile people with the Dutch healthcare services begins with a consideration of the significance and impact of religion. We then highlight the consequences of linguistic barriers between patients and providers and discuss the feelings of disrespect and discrimination reported by our participants. The chapter then focuses on the issue of travelling 'home' for treatment, as this was a further common theme in both studies.

Infertility and religion

Although Islam strongly encourages seeking biomedical solutions to reproductive challenges, not all treatments are allowed within Sunni Islam. There are several authoritative religious decrees, called fatwas, regarding assisted reproductive technologies (ARTs), which overrule several possibilities offered within the Dutch healthcare system. For example, ARTs involving donated gametes are not permitted in Sunni Islam (Serour 1996; Meirow and Schenker, 1997; Inhorn, 2006).[4] Also Islam does not favour adoption (although adoption is allowed by law in Turkey, which is a secular state).

When discussing the experiences of Turkish and Moroccan women, the influence of religion was evident in both studies and impacted on treatment experiences in several ways. A minority of participants had doubts about the acceptability of physical examinations, and reported that they had consulted Imams about this:

> *In the beginning he [husband] thought internal examinations were* Haram *[forbidden by religion], but after he visited the imam had agreed with everything*
>
> (Moroccan woman, Study One)

Also, a substantial number of the Moroccan participants travelled to Morocco to get authorization from religious personnel that particular infertility treatments, like IVF, were allowed by their religion. The Moroccan people in Study One were more likely to mention religious authorization for treatments as a reason for travelling to their country of origin than Turkish minorities in both studies. This may be because of their slightly more orthodox religious beliefs (SCP/WODC/CBS, 2005). A significant number (82 per cent) of Moroccan men considered their faith as important for decisions regarding their fertility treatment, while 64 per cent of the Turkish men said their belief had no influence upon their treatment (Korfker et al, 2008). However, the absence of a need to consult religious authorities among the Turkish population might also be explained by the fact that Turkish migrants are better informed and more familiar with which treatments are allowed by Islam because of the growing attention paid to infertility treatment in Turkey, within the Turkish media (van Rooij, 2008, Gürtin-Broadbent, Chapter 7).

In relation to religious prohibition, the negative stance of Sunni Islam towards gamete donation and surrogacy (Meirow and Schenker, 1997; Inhorn, 2006) was reflected in most participants' accounts:

> *I am afraid of semen from another person. It is* Haram.
>
> (Moroccan woman, Study One)

However, for the Turkish participants at least, issues relating to the absence of a genetic tie, emotional issues or the difficulty of finding a suitable donor also played a role in the refusal of treatments involving third party gametes or surrogacy (van Rooij, 2008). In this respect the participants in our studies did not differ from many other infertile couples (e.g. Greil, 1997; Kentenich and Yüksel, 1997; Becker, 2000).

In Study Two, several people expressed dissatisfaction with the way in which doctors proposed the use of semen or egg donation. Some reported that doctors should have more knowledge about, and show more respect for, the fact that this option is not allowed within their religion. The following account, however, shows that it is also important how this knowledge is used:

> *Yes, after three months I got the results by telephone. At my work. And he said there is zero percent chance to become pregnant. And then I thought, what is he saying? I said, but you would help me. He said, there is nothing to help you with, there is nothing, you entered early menopause. I said, how is this possible, it has to be possible. He said, what is your religion? I think so. I said Muslim. He said, you could use someone's else's egg to become pregnant, but you are a Muslim, then it is not possible. That was it, the conversation.*

(Turkish woman, Study Two)

In this example, besides the problem of giving such a message by telephone, the doctor does not allow the patient to make her own decision about whether or not to use donor eggs, but makes assumptions about her behaviour based on her stated religious beliefs. Although Sunni Islam does not allow egg donation, a woman might nevertheless decide to use this option in order to fulfill her desire for a child. In this example, the doctor denies the agency of the patient in her treatment seeking and effectively restricts her only chance of achieving a pregnancy. Special attention therefore, should be given to increasing the knowledge of healthcare providers about the cultural and religious background of their patients, but it is also important that providers sensitively explore options with patients and do not make assumptions about their choices on the basis of religious or other cultural affiliations. There are many people within all faiths that make choices outside official teachings and in line with their own conscience.

Several studies indicate that the Islamic religious beliefs of patients might also play a role in a preference for a female gynaecologist (e.g. Lafta, 2006; Yanikkerem et al, 2007). In Study One, the gender of healthcare staff appeared to be important to the Moroccan patients. Among Moroccan participants, a majority said that they would prefer or would only allow a gender-matched doctor. More than two-thirds of Moroccan women wanted a female doctor and more than half (58 per cent) of the Moroccan men had a strong preference for a male doctor for themselves. The majority of Moroccan men preferred a female doctor for their wife and two Moroccan men reported that they would refuse a male doctor for their wife. A considerable percentage of Turkish men (75 per cent) also preferred a female doctor for their wife. These preferences were unrelated to either knowledge of the Dutch language or length of stay in the Netherlands. Interestingly, only 38 per cent of the Turkish women preferred a female doctor for themselves in this study. Culley et al (2006) also found a preference for female practitioners among some infertile South Asian minority women based on cultural and religious grounds. However, it should be mentioned that Korfker et al did not ask the reason for this preference, so it might be that non-religious reasons also played a role. In the interviews in Study Two, the issue of gender-matched doctors was seldom raised by the Turkish minority participants, but it was also not explicitly asked by the interviewer. The Royal Dutch Society to Improve Health Care recently released a statement with respect to preferences for a male or a female doctor, stating that doctors should try to respect (if possible) the patient's wish in this respect, if it is requested within a reasonable time frame (e.g. when making an appointment) (http://knmg.artsennet.nl/content/resources//AMGATE_6059_100_TICH_R20 3395786518956//).

Linguistic barriers

Linguistic barriers in the communication with healthcare providers were a key theme which emerged in both studies. A majority of the Turkish and Moroccan respondents in both studies experienced such barriers in their communication with Dutch doctors and reported the negative impact of this language difference. Many mentioned the difficulties they had with understanding what the Dutch doctors said:

> *I did not understand everything. Sixty percent I understand, but talking is more difficult.*
>
> (Turkish man, Study Two)

> *We didn't understand the explanation of the doctor and had to go back to ask him again.*
>
> (Moroccan man, Study One)

Often one of the partners had a better knowledge of Dutch than the other. In Study One, Moroccan men were slightly more proficient in Dutch than Moroccan women. There was no discernable pattern relating to gender and language among the Turkish participants in either study. In the cases where one partner was more proficient at Dutch, doctors often expected partners to translate:

> *If he is present ... they know he hardly speaks Dutch, but they don't pay attention to this. They expect that the partner understands everything and will tell him.*
>
> (Turkish woman, Study Two)

This created additional problems for some patients. A few people, for example, reported that they did not trust the translations of their partners or that they felt that their partner was intentionally mistranslating information. In other cases people had a limited biomedical vocabulary and because of that were not able to translate information given by doctors. Additionally they were not able to translate written Dutch information to Turkish for their partners:

> *It was really difficult ... Sometimes, once in a while, they did give me some-*
> *thing with information on it. And then, then I had to translate that to*
> *Turkish, but it was not easy. Just sometimes, I'm not 100 per cent, mm,*
> *Turkish. I had to take a dictionary! Turkish–Dutch, Dutch–Turkish.*
>
> (Turkish woman, Study Two)

Culley et al (2006) also found that some patients who translated for their partners felt that having to both listen and understand, and then translate, caused
them an additional burden.

Having difficulties with translating and understanding biomedical information was also mentioned with respect to the content of written information:

> *We find it difficult; while my wife has good reading skills, we can't under-*
> *stand the content.*
>
> (Turkish man, Study One)

In addition to having a limited biomedical vocabulary, some people in Study
Two mentioned that they were able to speak Dutch, but had difficulties with
expressing emotions and pain to their doctors in Dutch as effectively as they
would be able to do in their own language.

These difficulties in communication meant that a significant proportion of
the participants in both studies did not feel fully informed by their Dutch
doctors about their infertility, examinations, treatments, possible side effects and
other possible treatment options. Communication problems between ethnic
minority patients and Dutch doctors, and the consequences of those problems,
have been studied with respect to other health issues. Poor communication leads
to feelings of confusion, increased chances of non-compliance, feelings of fear
and despair, problems in achieving rapport with care providers and dissatisfaction (Flores, 2005; Schouten and Meeuwesen, 2006).

With respect to infertility care of patients who are not fluent in Dutch, it
would be useful to use a trained interpreter (in person or by phone) or
develop information booklets that are parallel in Dutch and Turkish (see also
Culley et al, 2006) to enhance patients' understanding of examinations,
causes, treatment options and possible side effects. This might also enhance
shared decision making and meaningful informed consent. In Study One,
Korfker and colleagues had positive experiences working with migrant
outreach workers. They concluded that such workers, who share the language
and understand the cultural background of patients *and* healthcare providers,
can perform an important bridging function between migrant patients and
doctors. Joosten-van Zwanenburg et al (2004) also found that in the
Netherlands migrant outreach workers have a positive influence on
communication in the biomedical system.

Feelings of disrespect and discrimination

The findings from our research illustrate two further types of negative experiences with treatment: feelings of being treated with disrespect, and

discriminatory practices. In Study One, 10 per cent of the Moroccan women and 19 per cent of the Turkish women did not feel that they had been treated with respect:

> ... *while I was laying with open legs the female doctor suddenly introduced a male assistant without asking me.*
>
> (Moroccan woman, Study One)

Atkin (Chapter 3) points out the existence of institutional discrimination within healthcare services in the UK. Aspects such as not providing appropriate language support and prejudiced assumptions about ethnic minorities are mentioned as evidence of institutional racism. Some of these practices were also reported by the Turkish and Moroccan participants in the two studies presented here. Several patients in Study Two reported feeling that Dutch doctors are prejudiced and judged all Turkish migrants as being the same:

> *They look different at you. I think they have this idea, for instance: Oh look there you have yet another foreigner who can't have a baby and they immediately think that the man thinks that it will be the woman having a problem. You know. Prejudices.*
>
> (Turkish woman, Study Two)

Some people also felt that they were not being taken seriously because of their minority background:

> ... *maybe they behave like this because I am Turkish.*
>
> (Turkish women, Study Two)

In Study Two, some of the respondents in their early 20s also mentioned that Dutch doctors did not take them seriously because of the relatively young age (according to the perception of Dutch doctors) at which they were pursuing parenthood.

A small number of participants in Study Two also mentioned that they had at one point in their treatment believed they were discriminated against because of their ethnicity:

> *It felt like. Because we are Turkish, there are enough children in the Netherlands ... I thought they might let us waste three four years. So there won't be a child.*
>
> (Turkish woman, Study Two)

Some time later this participant found out that Dutch people also had to wait for certain treatments and realized that being placed on a waiting list was not in fact discriminatory. Such feelings of discrimination, however, are likely to have affected her relationship with her doctors and would not have occurred if she had been better informed (by providers), at the outset of treatment, about the procedures in the Netherlands.

The final theme presented here relates to one possible outcome of poor communication and institutional racism within the Dutch medical system; the practice of travelling abroad for infertility treatment.

Travelling 'home' for treatment

In addition to making use of the Dutch healthcare system, a substantial proportion of Turkish and Moroccan patients also visited infertility services in their country of origin. This was the case for 55 per cent of the Turkish patients and 25 per cent of the Moroccan patients in Study One and 45 per cent of the Turkish patients in Study Two. The lower percentage of Moroccans exercising this choice is probably a consequence of the better organization and availability of infertility services in Turkey than in Morocco. However, with the fast developing healthcare system in Morocco, the number of Moroccan minorities visiting infertility services in Morocco is likely to increase. Participants gave several reasons for deciding to travel 'home' for treatment. Many felt that they would be better informed about their problem and about treatment options if this was available in their own language:

> *And the Doctor in Turkey helped me a lot. Over there I could learn everything in my own language, the real meaning about what was the matter. Because lots of Dutch words I don't know. How hard I try, I think it is something which has to happen from birth onwards. Over there I learned everything and I also looked at internet.*
>
> (Turkish woman, Study Two)

In Study One a significant percentage of patients expressed concern about their treatment. Almost a third of the Turkish women and 15 per cent of the Moroccans reported that they had no confidence in the treatment they were receiving and a further 19 per cent expressed some doubts about their treatment:

> *… because I feel the doctor does not pay attention to my problems.*
>
> (Turkish woman, Study One)

Of those who had travelled home for treatment, many were critical of the Dutch medical system. For instance, aspects mentioned in both studies were: that they found the care impersonal; they received too little attention; there was a lack of time with doctors; too many different doctors were involved in their care; negative experiences related to examinations and forms of treatment; absence of second opinions; and high dissatisfaction with the speed with which they were treated. Long waiting lists, long intervals between treatments and between examinations, and delays in being referred to specialists were frequently mentioned:

> *… it's every two months a new investigation… In Turkey you can finish certain investigations within a month's time.*
>
> (Turkish woman, Study Two)

These concerns about the accessibility and quality of infertility services are also found among some Dutch patients (Kremer, 2007).

However, Turkish patients in Study Two, in particular, found that having contact with both the Turkish and Dutch biomedical systems sometimes caused confusion. Some couples used the Turkish system (leaflets, internet, media, consultations with doctors) to obtain necessary information about causes and possible treatments while undergoing treatment in the Netherlands. Differences between the two systems (attitudes, procedures and available treatments (Hunault et al, 2004; Karaman, 2005)) sometimes led to doubts about which procedures were 'correct' and a general dissatisfaction with the Dutch system.

Patients also reported negative stories about the Dutch system given to them by Turkish doctors and vice versa, and they also suggested that Dutch and Turkish doctors gave conflicting advice.

Travelling 'home' for treatment would appear to be a relatively common occurrence within minority ethnic communities, as reported in other studies (Yebei, 2000; Gacinski et al, 2002). However, it is important to note that indigenous Dutch people also sometimes utilize infertility services in other countries (Kremer, 2007). Reasons for such travel include a desire for treatments that are not available in the Netherlands (such as a wish to use anonymous gamete donors, prohibited in the Netherlands from 2004), the belief that their chances of pregnancy will be higher elsewhere, or because Dutch infertility clinics are reluctant to offer treatment if there is a very low chance of pregnancy or a high chance of a spontaneous conception (Kremer, 2007).

Conclusion

This chapter has shown that a substantial proportion of the Moroccan and Turkish men and women in our research experience significant problems within infertility care. Problems in their interactions with the Dutch healthcare system are likely to aggravate the considerable negative personal and social experiences they already have as a consequence of their involuntary childlessness (van Rooij et al, 2007; van Rooij, 2008). Some of the problems expressed by our participants are common among all patients of Dutch infertility services, regardless of ethnicity, in particular, the relatively slow nature of the Dutch infertility care system and the number of different doctors involved in patient care (Kremer, 2007; Korfker et al, 2008; see also Culley et al, 2006, for the UK). However, some negative aspects of care appear to be more applicable to ethnic minorities. This chapter illustrates that infertility services appear, to some extent, to be unable to respond to the needs of a multi-cultural Dutch society. Medical personnel appear to have difficulties with adequately communicating necessary information to non-Dutch speakers and they rarely provide language support. Some migrants were also confronted with prejudice. Furthermore, institutional barriers to effective care appear to contribute to poor access for minorities, some of whom feel that it is necessary to go to their country of origin for treatment and/or advice.

These problems require attention in the training of healthcare practitioners. More emphasis should be placed on how to work across cultural and linguistic boundaries. Furthermore, infertility services should consider the use of trained interpreters (in person or by phone) (see also Flores, 2005, for the positive effects of trained interpreters) and/or migrant outreach workers (Joosten-van Zwanenburg et al, 2004). Additionally, information booklets and audiovisual materials (see also Culley et al, 2004, 2006) that are parallel in Dutch and Turkish, Moroccan Arabic or Berber are needed. Better information is required about causes of infertility, examinations and treatments, as well as about the Dutch medical system and the likely timescale for treatment. At the same time, migrants should be better supported to learn Dutch to facilitate their ability to negotiate the healthcare system. Such changes would result in minorities receiving a higher quality of care, feeling more satisfied with their treatment, being better able to share in decision making and exercising informed consent. It might also mean that fewer people would need to use infertility services in their country of origin.

Notes

1   According to official definitions of Statistics Netherlands (2008), 'non-Western' countries are countries in Africa, Latin America and Asia (with the exception of Indonesia and Japan). 'Western' countries involve all other countries. Non-Western migrants are defined as people who have at least one parent who was born in a non-Western country. In many parts of Europe the term 'migrant' is used to refer both to immigrants *and* those who are born in 'host' countries, but whose ancestral origins lie elsewhere. The term more commonly used in UK studies is 'minority ethnic'.

2   To date, more than 19 per cent of the population of the Netherlands is of non-Dutch origin: 8 per cent is of Western origin and 11 per cent of non-Western origin (Statistics Netherlands, 2007). The largest groups of migrants of non-Western origin in the Netherlands are comprised of people with a Surinamese, the Netherlands Antillean, Turkish or Moroccan background. While Surinam was a Dutch colony, the Netherlands Antilles are still part of the Dutch state. Turkish and Moroccan people came mainly as labour migrants and, later, for marital purposes or to be reunited with their family (Hooghiemstra, 2003).

3   The Dutch system is a private health insurance system with social conditions, operated by several private health insurance companies, which are obliged to accept every resident in their area of activity. Everyone pays a nominal premium to the health insurer. The health insurance consists of a standard package of essential healthcare. Additionally, it is possible to pay for a wider coverage. Everyone with the same policy will pay the same insurance premium. In the Netherlands it is compulsory to have healthcare insurance (www.minvws.nl/en/themes/health-insurance-system/default.asp, accessed 26 September 2008). Nevertheless, in 2007 between 1 and 2 per cent of the people in the Netherlands were not insured (Statistics Netherlands, 2008). The standard package of healthcare insurances covers a substantial proportion of infertility treatments. The coverage of additional packages might vary between healthcare insurances. However, during the past few years there have been several changes in the standard coverage with respect to the number of IVF treatments attempted and medication for fertility treatments, due to governmental decisions (www.minvws.nl;

www.freya.nl). For instance, up to 2003, healthcare insurances provided for three attempts of IVF/ICSI per pregnancy. Between 1 January 2004 and 1 January 2006 insurance policies would cover only the second and third treatment. In 2007 this reverted to the previous situation where three IVF/ICSI treatment cycles per pregnancy are covered. Also, for several years some treatments were not covered (for instance, intrauterine insemination (IUI)). However, the government is currently reconsidering the coverage of these treatments. Nevertheless, infertility clinics sometimes refuse prospective single woman, women of a relatively old age (over 40–45), lesbian couples, older men or carriers of a genetic disease (Hunfeld et al, 2004).

4   While Sunni Islam has a negative stance towards gamete donation, it is not so clear cut in the case of the Shia (see Inhorn, 2006).

References

Balen, F. van and Inhorn, M. C. (2002) 'Interpreting infertility: A view from the social sciences', in M. C. Inhorn and F. van Balen (eds) *Interpreting Infertility: Childlessness, Gender, and Reproductive Technologies in Global Perspective*, UCLA-press, Berkeley, CA

Becker, G. (2000) *The Elusive Embryo: How Women and Men Approach New Reproductive Technologies*, University of California Press, Berkeley, CA

Culley, L. A., Rapport, F., Katbamna, S., Johnson, M. and Hudson, N. (2004) *A Study of the Provision of Infertility Services to South Asian Communities*, De Montfort University, Leicester

Culley, L. A., Hudson, N., Rapport, F. L., Katbamna, S. and Johnson, M. R. (2006) 'British South Asian communities and infertility services', *Human Fertility*, vol 9, no 1, pp37–45

Daar, A. S. and Merali, Z. (2001) 'Infertility and social suffering: The case of ART in developing countries', in E. Vayena, P. J. Rowe and P. D. Griffin (eds) *Current Practices and Controversies in Assisted Reproduction*, World Health Organization, Geneva, pp15–21

Dyer, S. J. (2007) 'The value of children in African countries: Insights from studies on infertility', *Journal of Psychosomatic Obstetrics and Gynecology*, vol 28, no 2, pp69–77

Evenblij, M., Mackenbach, J. and van der Veen, E. (2004) *Gezondheid in Kleur: Nieuwe Inzichten uit het Onderzoeksprogramma Cultuur & Gezondheid [Health in Colour: New Insights from the Research Programme Culture & Health]*, Aksant, Amsterdam

Ferguson, W. J. and Candib, L. M. (2002) 'Culture, language, and the doctor–patient relationship', *Family Medicine*, vol 34, no 5, pp353–361

Flores, G. (2005) 'The impact of medical interpreter services on the quality of health care: A systematic review', *Medical Care Research and Review*, vol 62, no 3, pp255–299

Gacinski, L., Yüksel, E. and Kentenich, H. (2002) 'The unfulfilled desire for a child of oriental couples: Infertility counseling and treatment for Turkish immigrants in Germany', in B. Strauss (ed) *Involuntary Childlessness, Psychological Assessment, Counseling and Psychotherapy*, Hogrefe & Huber Publishers, Seattle, WA

Gerrits, T. (1997) 'Social and cultural aspects of infertility in Mozambique', *Patient Education and Counseling*, vol 31, pp39–48

Greil, A. L. (1997) 'Infertility and psychological distress: A critical review of the literature', *Social Science & Medicine*, vol 45, pp1679–1704

Guntupalli, A. M. (2004) 'Perceptions, causes and consequences of infertility among the Chenchu tribe of India', *Journal of Reproductive and Infant Psychology*, vol 22, no 4, pp249–259

Harmsen, J. A. M., Bernsen, R. M. D., Bruijnzeels, M. A. and Meeuwesen, L. (2008) 'Patients' evaluation of quality of care in general practice: What are the cultural and linguistic barriers?', *Patient Education and Counseling*, vol 72, no 1, pp155–162

Heelsum, A. van (2003) 'Moroccan Berbers in Europe, the US and Africa and the concept of diaspora', *International Studies Electronic Publishing Program* (of UC Press, the California Digital Library, and the international and area studies units on all UC campuses)

Hooghiemstra, E. (2003) *Trouwen over de Grens. Achtergronden van Partnerkeuze van Turken en Marokkanen in Nederland [Crossing Borders. Backgrounds of Partner Selection of Turkish and Moroccan People in the Netherlands]*, SCP, The Hague http://knmg.artsennet.nl/content/resources//AMGATE_6059_100_TICH_R203395786 518956//, accessed 15 August 2008

Huis, L. T. van (2008) 'Partnerkeuze van allochtonen [Partner choice of migrants]', *Demos*, vol 24, no 1, pp1–4.

Hunault, C. C., Habbema, J. D. F., Eijkemans, M. J. C., Collins, J. A., Evers, J. L. H. and te Velde, E. R. (2004) 'Two new prediction rules for spontaneous pregnancy leading to live birth among subfertile couples, based on the synthesis of three previous models', *Human Reproduction*, vol 19, no 9, pp2019–2026

Hunfeld, J. A. M., Passchier, J., Bolt, L. L. E. and Buijsen, M. A. J. M. (2004) 'Protect the child from being born: Arguments against IVF from heads of 13 licensed Dutch fertility centres, ethical and legal perspectives, *Journal of Reproductive and Infant Psychology*, vol 22, no 4, pp279–289

Husain, F. A. (2000) 'Reproductive issues from the Islamic perspective', *Human Fertility*, vol 3, pp124–128

Inhorn, M. C. (1996) *Infertility and Patriarchy: The Cultural Politics of Gender and Family Life in Egypt*, University of Pennsylvania Press, Philadelphia, PA

Inhorn, M. C. (2003) *Local Babies, Global Science: Gender, Religion, and In Vitro Fertilization in Egypt*, Routledge, London

Inhorn, M. C. (2006) 'Making Muslim babies: IVF and gamete donation in Sunni versus Shi'a Islam', *Culture, Medicine and Psychiatry*, vol 30, no 4, pp427–450

Inhorn, M. C. and Fakih, M. H. (2006) 'Arab Americans, African Americans, and infertility: Barriers to reproduction and medical care', *Fertility & Sterility*, vol 85, no 4, pp844–852

Inhorn, M. C. and van Balen, F. (2002) *Infertility around the Globe: New Thinking on Childlessness, Gender and Reproductive Technologies*, University of California Press, Berkeley, CA

Joosten-van Zwanenburg, E., Kocken, P. and de Hoop, T. (2004) *Onderzoek naar de Effectiviteit van de Inzet van Allochtone Zorgconsulenten in Rotterdamse Huisartsenpraktijken in de Zorg aan Vrouwen van Turkse en Marokkaanse Afkomst met Stressgerelateerde Pijnklachten [Study into the Effectiveness of Using Migrant Care Consultants in General Practitioner Practices in Rotterdam in the Care for Women with a Turkish and Moroccan Background with Stress Related Pain Symptoms]*, GGD Rotterdam en Omstreken, Rotterdam

Karaman, Y. (2005) Personal communication, 12 January 2005

Kaya, H. (2006) 'Door diversiteit in geloofsovertuigingen van bevolkingsgroepen, Verschillen in integratie van Turken in Nederland [Because of diversity in religion of population, differences in integration of Turkish people in the Netherlands]', *Demos*, vol 22, no 5, pp45–48

Kentenich, H. and Yüksel, E. (1997) *Psychosomatisches Betreuungskonzept Steriler Turkischer Paare in der Migration [Psychosomatic Concept of Care for Migrated Sterile Turkish Couples]*, Virchow-Klinikum, medizinische fakultat der HUB, Frauen- und

Poliklinik Frauen- und Kinderklinik, DRK-Klinikern Westend, Berlin

Korfker, D. G., van der Pal-de Bruin, K. M., Detmar, S. B. and Buitendijk, S. E. (2008) *Knelpunten bij de Behandeling van Infertiliteit van Turken en Marokkanen [Problems in Infertility Treatment of Turkish and Moroccan people]*, TNO-KvL/P&Z 2008.056, TNO, Leiden

Kremer, J. (2007) 'Zorg over de grens [Care across borders]', *Medisch Contact*, vol 33/34, pp1343–1346

Lafta, R. K. (2006) 'Practitioner gender preference among gynecologic patients in Iraq', *Healthcare for Women International*, vol 25, pp125–130

Lodewijkckx, E. and Hendrickx, K. (1996) 'Alleen al de angst om de pil te vergeten maakt mij ziek: Een kwalitatief onderzoek bij Marokkaanse vrouwen [Only the anxiety about forgetting the pill makes me sick: A qualitative research among Moroccan women]' *Bevolking en Gezin*, vol 2, pp61–86

Meeuwesen, L., Harmsen, J. A. M., Bernsen, R. M. D. and Bruijnzeels, M. A. (2006) 'Do Dutch doctors communicate differently with immigrant patients than with Dutch patients?', *Social Science & Medicine*, vol 63, pp2407–2417

Meirow, D. and Schenker, J. G. (1997) 'The current status of sperm donation in assisted reproduction', *Journal of Assisted Reproduction and Genetics*, vol 14, no 3, pp133–138

Nahar, P. (2007) 'Childless in Bangladesh', PhD thesis, University of Amsterdam, Amsterdam

Nazroo, J. Y. (2003) 'The structuring of ethnic inequalities in health: Economic position, racial discrimination, and racism', *American Journal of Public Health*, vol 93, no 2, pp277–284

Rooij, F. B. van (2008) 'The experience of involuntarily childless Turkish migrants in the Netherlands: Parenthood motives, psycho-social consequences, responses and help-seeking behavior', PhD thesis, SCO Kohnstamm instituut, University of Amsterdam, Amsterdam, www.freya.nl, accessed 15 August 2008

Rooij, F. B., van Balen, F. and Hermanns, J. M. A. (2006) 'Migrants and the meaning of parenthood: Involuntary childless Turkish migrants in the Netherlands', *Human Reproduction*, vol 21, no 7, pp1832–1838

Rooij, F. B. van, van Balen, F. and Hermanns, J. M. A. (2007) 'Emotional distress and infertility: Turkish migrant couples compared to Dutch couples and couples in Western Turkey', *Journal of Psychosomatic Obstetrics & Gynaecology*, vol 28, no 2, pp87–95

Schouten, B. C. and Meeuwesen, L. (2006) 'Cultural differences in medical communication: A review of the literature', *Patient Education and Counseling*, vol 64, nos 1–3, pp21–34

SCP/WODC/CBS (2005) *Jaarrapport Integratie 2005 [Annual Report on Integration 2005]*, The Hague

Serour, G. I. (1996) 'Bioethics in reproductive health: A Muslim perspective', *Middle East Fertility Society Journal*, vol 1, pp30–35

Serour, G. I. (1998) 'Reproductive choice: A muslim perspective', in J. Harris and S. Holm (eds) *The future of Human Reproduction: Ethics, Choice and Regulation*, Oxford University Press, New York

Serour, G. I. (2001) 'Attitude and cultural perspectives on infertility and its alleviation in the Middle East area', in E. Vayena, P. J. Rowe and P. D. Griffin (eds) *Current Practices and Controversies in Assisted Reproduction*, World Health Organization, Geneva

Statistics Netherlands (2007) StatLine [Data file], available at http://statline.cbs.nl/statweb

Statistics Netherlands (2008) StatLine [Data file], available at http://statline.cbs.nl/statweb

Uiters, A. H. (2007) 'Primary health care use among ethnic minorities in the Netherlands: A comparative study', PhD thesis, Erasmus University, Rotterdam

Uiters, A. H., Devillé, W. L. J. M., Foets, M. and Groenewegen, P. P. (2006) 'Use of health care services by ethnic minorities in The Netherlands: Do patterns differ?', *European Journal of Public Health*, vol 16, no 4, pp388–393

Valk, H. A. G. de (2001) '"Ze zeiden dat het een paradijs was hier ..."; migratie en de rol van informatie ["They told us it was a paradise here ..."; migration and the role of information]', *Demos*, vol 17, no 2, pp9–11

Yanikkerem, E., Ozdemir, M., Bingol, H., Tatar, A. and Karadeniz, G. (2007) 'Women's attitudes and expectations regarding gynaecological examination, *Midwifery*, doi: 10.1016/j.midw.2007.08.006

Yebei, V. N. (2000) 'Unmet needs, beliefs and treatment-seeking for infertility among migrant Ghanaian women in the Netherlands', *Reproductive Health Matters*, vol 8, no 16, pp134–141

Yüksel, E. (1995) 'Kinderlosigkeit bei Turkischen Paaren: "Wie ein Baum ohne Fruchte" [Childlessness among Turkish couples: "Like a tree without fruit"]', *TW Gynakolongie*, vol 8, no 6, pp461–467

9

Treating the Afflicted Body: Perceptions of Infertility and Ethnomedicine among Fertile Hmong Women in Australia

Pranee Liamputtong

Introduction

In this chapter, I explore the cultural interpretation of fertility and infertility among fertile Hmong women from Laos who are now living in Australia. In the Hmong world, having children is essential for parents' well-being, not only in this world but also for the next. Children, particularly sons, carry on the family lineage, look after the elderly parents when they are alive, and worship them when they die. Without sons, deceased parents will not be reincarnated. Without children in the lineage and clans, Hmong society will cease its existence. Fertility is, therefore, a most crucial part of Hmong life. The failure to be fertile has a profound effect on Hmong lineages, clans and society (Symonds, 1991, 1996, 2004; Liamputtong Rice, 2000; Liamputtong and Spitzer, 2007).

Infertility also taints women's moral identities. As in many other societies, infertility is seen as unnatural and it is the woman who bears the blame (see for example, Inhorn, 1994a; Neff, 1994; Kielmann, 1998; Prapreen et al, 2000; Riessman, 2000; Matsubayashi et al, 2001; Bharadwaj, 2003; Guntupalli and Chenchelgudem, 2004; Mariano, 2004; Johnson-Hanks, 2006; Donkor and Sandall, 2007; Rashid, 2007; Hollos and Larsen, 2008). I shall demonstrate that in Hmong culture an infertility problem is perceived to be located within the woman's body. The ethnomedical reasons for her 'afflicted body' are many, and treatments are to be found in both the natural and supernatural realms.

This chapter is based on an ethnographic study with Hmong women in Melbourne, Victoria. The study covered a number of issues concerning reproductive health, including the beliefs and practices related to fertility and infertility. It must be noted that most Hmong women in this study were fertile.[1] Yet issues relating to infertility and its treatment were discussed at length when I asked women about conception, pregnancy, birth and motherhood. This chapter is constructed from women's discourse on what Inhorn (1994a) has called the 'fertility–infertility dialectic' – how discussion of fertility may lead to the discussion of infertility and vice versa.

The majority of the women had experienced childbirth while living in Laos or in a refugee camp in Thailand as well as in hospitals in Melbourne. In addition, I interviewed several Hmong traditional healers (shamans, medicine women and a magic man) who are now living in Melbourne, in order to obtain more in-depth detail of traditional healing methods. All interviews were conducted in the Hmong language with the assistance of a bicultural research assistant who is a Hmong native-born woman.

In addition, participant observation allowed me to observe and record Hmong cultural beliefs and practices, as experienced in Australia, more fully. I attended a number of Hmong rituals and ceremonies, such as soul-calling ceremonies and shamanic rituals performed by shamans, and participated in Hmong activities including walking with a medicine woman in her herbal garden and attending the Hmong New Year. The main interviews and participant observation were conducted between May 1993 and July 1994. However, since the initial fieldwork, I also made several returns to the community in order to obtain more up-to-date information as well as to keep contact with my key informants.

The Hmong: Social and cultural backgrounds

The Hmong in Australia come from Laos where they lived as hill tribes in the high mountainous areas. Involved in the fighting between the American forces and the Pathet Lao, the communist group in Laos, the Hmong were forced to move out of their homeland in the mountains and escape to Thailand (Quincy, 1995; Fadiman, 1997; Cha, 2000, 2003; Culhane-Pera et al, 2004). The US accepted the majority as migrants. It has been estimated that there are between 227,217 and 268,747 Hmong people living in America at present (Hmong National Development Links, 2001, cited in Culhane-Pera et al, 2004). In Australia, the number is far smaller than in the US. It is estimated that there are about 1800 Hmong in Australia (Lee, 2004; Tapp, 2004). This number is based on an estimation of Hmong community leaders in Melbourne only, as there are no official statistics collected on the Hmong per se. The Hmong in Australia are included under 'people born in Laos'.

The main concentration of the Hmong in Australia nowadays is in far north Queensland, particularly in Cairns, Innisfail and Atherton. But, many Hmong are still living in Victoria, New South Wales and Tasmania (Lee, 2004; Tapp and Lee, 2004). In Victoria, where I carried out my study with Hmong people, the Hmong live in close-knit groups, mainly in high-rise public housing in one of the inner suburbs of Melbourne, and there is another group in an outer suburb (Liamputtong Rice, 2000; Liamputtong, 2004).

The Hmong in my study are much poorer than other Southeast Asian refugees. The majority of Hmong people are unemployed, lack formal education and are still learning English. The Hmong are animistic and follow ancestor worship. They believe in reincarnation; the rebirth cycle. The Hmong are patrilineal and patrilocal. Family names follow the clan system (Cooper et al, 1995; Cha, 2003; Symonds, 2004). There are ten clans in Melbourne. The usual Hmong family is large. Most Hmong women in the study described here have between four and six children and it is likely that they will continue to have more children. Traditionally, the Hmong put a high value on having many children, particularly boys, since they could help in farming and continue traditional practices such as worshipping ancestral spirits, caring for their parents in old age, and carrying on the clan name (Symonds, 1991, 2004; Cooper et al, 1995; Liamputtong Rice, 2000; Cha, 2003; Symonds, 2004). Such traditional customs continue to be practised in Australia.

Infertility and woman's identity

In the Hmong culture, children are regarded as the prosperity of the family; therefore having children is highly desirable. Hmong women are expected to be able to produce many children. Ideally, to be a good woman she must bear at least as many children as her mother or mother-in-law (Liamputtong Rice, 1997, 2000; Liamputtong and Spitzer, 2007).

Why are children important and so valued? From a 'social structural perspective', Inhorn (1996, p247) argues that children are important for three reasons. First, children are necessary in order to secure the survival of their parents and families. Second, in most patriarchal societies children are the only 'valuable power resource' for their mothers. Hence, having children can improve the status of the woman. Third, children are needed for the continuation of society (Liamputtong, 2007; see also Gyekye, 1996; Kimani and Olenja, 2001; Inhorn and van Balen, 2002; Hollos and Larsen, 2008).

In a similar vein, in Hmong society, children are necessary for one's wellbeing, not only in this life but in the afterlife. If a couple does not have sons, or daughters to marry out, there will be no one to look after them in old age. Sons are more important than daughters since they care for the family altar, feed the ancestors and carry on the '*dab qhuas*' (the clan spirit) to the next generation. This ensures the continuation of the family, lineage and clan (Liamputtong, 2007; see also Bharadwaj, 2003; Inhorn and van Balen, 2002).

While Hmong girls and adolescents do not have equal social or familial status to Hmong boys and men, their status changes when they marry and are able to bear a child (Symonds, 2004; Liamputtong Rice, 2000). The birth of the first child brings prestige to a Hmong woman. After giving birth to her first child a Hmong woman becomes known by the name of the child as its mother, for example 'Blia's mother'. She is no longer addressed as 'my younger daughter' by her parents, nor as the wife of her husband (as in 'Mrs Lee Xiong') by others (Symonds, 1991). Women gain respect and status when they produce children (Liamputtong Rice, 1995, 1997, 2000; Kimani and Olenja, 2001; Johnson-Hanks, 2006).

A woman who cannot bear children in the Hmong culture is known as one who has a difficult life. She is always seen as unfortunate, unlucky and unfulfilled, not only in this life but also in the next (Liamputtong and Spitzer, 2007). She is also not regarded highly in the community. Given these beliefs, it is not surprising that Hmong women experience pressure to produce children (Liamputtong Rice, 2000; Liamputtong and Spitzer, 2007). In Hmong culture polygyny is usually practised if a first wife is unable to produce children. Thus, if a woman remains infertile after a few years of marriage the husband is encouraged to take a second wife (cf. Okonofua et al, 1997; Kielmann, 1998; Koster-Oyekan, 1999; Hollos and Larsen, 2008). This encouragement usually comes from the man's parents. They look for another young and physically strong woman for their son.

The first wife has no say in this since she is unable to bear a child, and she becomes the 'big sister' among her husband's wives. If the other wives are able to produce children, she becomes the 'oldest mother' to her husband's children. Her status may not be affected if the junior wives and their children respect her and treat her as their 'big mother'. Problems arise only when she cannot get along well with the junior wives and her husband takes sides with his new wives. In some cases, the woman may be divorced and forced to move out of her husband's home. Most Hmong women who I interviewed do not wish to become a senior wife because of the insecurity of this position. Many say that their husband will eventually favour a wife who can bear him children, and they will be seen as worthless.

Another more important issue is the need to have a son in the family (Bharadwaj, 2003; Liamputtong, 2007). The Hmong are patrilineal, and the eldest son takes care of elderly parents. Upon the death of the parents, the son must honour and feed his deceased parents' spirits. This is to ensure that the parents' souls can reach the ancestral pool in the other world and be able to reincarnate. Sons are crucial for this important function as women are not allowed to perform ancestral worship. If a woman is unable to bear a son, she is again obliged to let her husband bring in another wife (Liamputtong Rice, 2000). Becoming a mother and having children also means that women will not be too lonely in old age since there will be many grandchildren around.

'Ethnogynecological causes' of infertility

Hmong women provide several layers of explanations about the causes of infertility. Their explanations about 'ethnogynecological causes', to use Inhorn's term (1994a, p163), follow the analytical framework proposed by Inhorn (1994a, p10) in her writing about infertility in Egypt.

Drawing on the work of Lock and Scheper-Hughes (1990), Inhorn (1994a) argues that when one attempts to discern causal explanations of infertility, there are three levels of 'causal proximity': (1) the 'proximate causes' located within 'the individual body'; (2) the 'medial causes' located within the 'social body' and involving social relations and the results of actions between the women and others including those of supernatural beings; and (3) the 'ultimate proximity' located within 'the body politic', and often involving divine will. Inhorn (1994a, p165) maintains that conceptualizing women's explanations of infertility in terms of 'causal proximity' provides readers with a more accurate picture of the ways women think about causality. In this section, I will present my findings accordingly.

The individual body: The cold womb

The Hmong have long associated physical health and well-being with bodily equilibrium (Liamputtong Rice, 2000; Vang et al, 2002; Cha, 2003; Culhane-Pera et al, 2004; Symonds, 2004). This belief derives from traditional Chinese medicine which asserts that an individual body is possessed of two opposite qualities: hot and cold. The individual body must be kept in a harmonious state with respect to these two qualities in order to be healthy. Ill health occurs when this harmony is disrupted. This belief is strongly maintained in health practices relating to women's reproduction (Symonds, 1996, 2004; Liamputtong Rice, 2000). During the postpartum period, for example, women avoid consuming cold food, taking cold showers and washing their hair since their bodies are believed to be in a cold state. The Hmong believe that a woman who has just given birth has been depleted of heat. She hence must observe certain restrictions to regain lost heat. Failure to observe such proscriptions will result in illness in later life. Infertility is no exception. The Hmong believe that a woman may be infertile because her womb is in a state of being 'cold'. Because the womb is cold, a foetus cannot be conceived. Women say that a cold womb can be caused by consuming cold foodstuffs during menstruation. Since menstruating women are in a state of coldness, the cold food will congeal menstrual blood and make the womb cold and this causes infertility.

The social body

The menstruating body

Given the association between blood and conception, it is not surprising that infertility is believed to be related to menstruation. A woman may become infertile due to her poor menstruation. The Hmong make a distinction between good and bad menstruation (Liamputtong, 2004). Colour of the menstrual blood is used as an indicator of good or bad blood. It is believed that good menstrual blood must be bright red. If it is dark, it is 'bad'. Bright blood indicates that the woman is healthy. In contrast, dark blood is found among pale and skinny women and is an indication of bad health. The flow of blood is also used to identify good or bad menstruation. Moderate flow is believed to be best for any woman. A combination of light flow and dark colour indicates bad health. The humoral quality of menstruation is closely related to fertility. It is believed that 'cold menstruation' contributes to infertility.

Good menstruation, without problems such as back and abdominal pains, means good conception. In other words, the colour, the flow and the humoral quality of menstruation are all indicators of the capability to conceive.

Menstruating women are subjected to some restrictions. For example, women should not perform heavy work and rigorous physical activities. This is believed to cause the uterus to 'turn' into a wrong position, and this results in infertility. In addition, women should not eat 'cold' food or drink icy cold water while menstruating. The cold foodstuff will congeal menstrual blood and make the womb cold, and this causes infertility as mentioned in the last section.

However, the most common explanation and most feared cause of infertility among the Hmong is '*raug dab*' (struck by spirits). In the Hmong world there are 'spirits of ancestors, sky and earth, food and water, sun and moon, and wind' (Thao, 1984, p327). Spirits are everywhere and they do not wish to be disturbed. If people do not keep away from the spirits they may be struck and become ill (Chindasri, 1976; Cooper et al, 1995; Symonds, 1996, 2004; Culhane-Pera et al, 2004; see also Chapman, 2007; Rashid, 2007).

In the case of infertility, a young woman may have crossed rivers or streams while menstruating and washed the clothes worn while menstruating in these rivers or streams. This activity is believed to disturb and anger the spirits of the rivers or streams. Hmong women are seriously warned against these activities. The spirits are able to trace a woman through the smell of her menstruation and 'strike' her. This causes miscarriage whenever she conceives. Repeated miscarriages in turn cause infertility in the woman.

A woman can become infertile, after having a first child, if the childbirth blood is washed away in the stream. The blood is again believed to cause the anger of the spirits. The woman is then struck by the spirits and she becomes infertile.

Women who have bad menstrual pain will also be infertile, as menstrual pain is an indication that a woman has invoked the spirit's anger by not observing a strict rule while menstruating. The pain is due to the fact that the spirits 'throw rocks' at her body. Because of the repeated injury of her body she is unable to conceive a child.

The badly behaved body

A woman's bad behaviour toward others in the society can also make her infertile. This includes saying bad words about other people, particularly young children, or making comments about other people's physical appearance causing anger among those insulted. Those so offended may make a curse on the insulter's body. The curse, which does not involve recourse to a sorcerer, is believed to be powerful enough to cause infertility in the woman who misbehaves.

A woman's bad behaviour toward others also invokes deities' anger since the deities see everything from the sky and keep a record of humans' behaviour. The deities consider a woman like this as not suitable to be a parent, and so will not send children to her. As we will see in the next section, without consent of the deities, the woman will remain infertile.

The body politic: Mandate of life

The Hmong also believe that an infertile woman may not have brought 'a piece of paper'[2] which indicated that she would have children from heaven when she was born. This piece of paper is written and given out by '*txoov kab yeeb*', a spirit couple who are believed to assist in fertility. The Hmong understand that to be able to conceive, a woman must have a husband or at least a man involved in the affair. However, pregnancy cannot happen solely by just having a husband. It is determined by *txoov kab yeeb*: a child is born only when *txoov kab yeeb* give permission which they state in her 'piece of paper' before her own birth.

It is also said that a child cannot be born because '*choj ntub*', the path for the child to travel from heaven, has been broken. The broken path prevents the child from finding his or her parents' home; therefore, (s)he cannot be born:

[Choj ntub]... *is there so the baby will have a path to come and she [the woman] will be able to have children. If it is not done like this then ... it is like a jungle if there is no path and you were in it then where do you go? Like the river too if there is no bridge how do you go across? If you go then you will drown in the river and so you won't be able to go unless there is a bridge.* (Mai)

'Ethnogynecologists' and the healing processes

Given the various ethnomedical causes of infertility described here, it is not surprising that infertile Hmong women seek help from ethnomedical specialists. There are at least two groups of 'ethnogynecologists' (Inhorn, 1994a, p97) from whom women seek help with infertility problems. Initially, a woman tries to resolve the problem alone by seeking help from '*kws tshuaj*' (a medicine woman). If the failure to conceive continues, her husband and the extended family will become involved, and normally '*txiv neeb*' (a shaman) will be consulted.

Kws tshuaj (a medicine woman)

Hmong women tend to seek help from a local medicine woman referred to as *kws tshuaj*, as a first line of resort. According to Culhane-Pera and colleagues (2004, p732), *kws tshuaj* are often women who obtain their knowledge by being an apprentice of an older female relative. *Kws tshuaj* 'are guided in diagnosing disease and prescribing medicines by their helping spirits (*dab tshuaj*)'. *Kws tshuaj* usually has her altar for her *dab tshuaj* inside the house, which is located next to the main household altar for the house spirit (*dab xwm kab*). In Melbourne, there was only one medicine woman who still practised at the time that I undertook this study, and Hmong women and families continued to seek help from her.

When they go to ask for help they must bring a bunch of incense sticks tied with a piece of red material, some gold money and several silver coins. These will be put on the medicine woman's altar. The woman then can ask for help. The medicine woman will examine her womb by massaging her stomach in a circular motion. From this, she is able to determine the condition of the woman's womb. The medicine woman then goes out to her garden and picks up several kinds of '*tshuaj Hmoob*' (Hmong green herbs). She will instruct the woman how to take the herbs. Usually, these are steamed with eggs and then eaten. This is done in the morning over three consecutive days. The woman then must wait for one year to see if the treatment is successful. During the one year period, the woman is not allowed to seek help from other medicine women; only if after the one year trial the treatment has not been successful, may the woman consult another medicine woman. If the treatment works and the woman becomes pregnant, then she must honour the medicine woman who helped her. Traditionally, this is done on the third morning after the woman has given birth, when the soul calling and naming ceremonies for the newborn infant are conducted. During the ceremonies the new mother must acknowledge the medicine woman's power and kneel down to thank her. The family must also prepare a chicken for her to bring home and they must pay her several silver coins. Some of the coins will be placed on the medicine woman's altar. This is to honour the spirit who assists in the medicine woman's successful intervention.

Txiv neeb (a shaman)

Shamans are an important part of Hmong life (Cha, 2000, 2003; Culhane-Pera et al, 2004; Symonds, 2004; Liamputtong, 2007). In Laos (and elsewhere, including Australia) there must be at least one shaman in every village or community. Without the shaman, it is said that Hmong do not feel 'secure' and the village may not continue to exist (Chindasri, 1992). In Australia, there is at least one shaman in each state, and there are at least four shamans in Melbourne.

The rituals of a shaman are mainly concerned with fertility, protection and curing (Cha, 2000, 2003; Culhane-Pera et al, 2004; Liamputtong, 2007). To ask a shaman to assist in fertility, a childless couple must go to the shaman's house where his altar is located, and request help formally. Once this is done, the shaman first holds a ritual '*ua neeb saib*' to determine the factor that contributes to the woman's childlessness, ascertaining whether she does not have 'a piece of paper', if she has been '*raug dab*' (struck by spirits), or she has disrupted social relationships in the society. If it is the first, and the woman did not bring her paper with her, then the shaman holds a ritual '*ua neeb cais*' to invoke '*txoov kab yeeb*' to aid in fertility. The shaman must offer *txoov kab yeeb* a sacrificed chicken and some gold money as a reward for changing the woman's paper and for bringing children to her. In his trance, with the assistance of his helper, a shaman will put '*daim nyias*' (a traditional Hmong baby back carrier) on his back and travel to beg *txoov kab yeeb* to send children to the couple. If the wish is granted, then the *daim nyias* symbolically carries the child down to earth.[3]

In the case of '*choj ntub*' (a broken path), a shaman has to build a bridge or path for the child to travel to his parents' home. A roll of white material (mostly white cloth) is rolled from the front door to the bed of the couple. The shaman, in his trance of performing '*ua neeb kho*' has the *daim nyias* wrapped on his back, travels to heaven and negotiates with *txoov kab yeeb* to give the couple children. If this request is granted, the shaman will carry the child in the *nyias* and travel down through the white material. In the shamanic performance, this is done by pulling the material from the front door into the bedroom, so that the child will have a path to follow. It is believed that only through this ritual will a woman become pregnant if she has *choj ntub*.

In the case of being struck by spirits, the shaman also has to hold the ritual *ua neeb cais* to negotiate with the spirits and bribe them to leave the woman in peace, allowing her to conceive. Traditionally, the spirits will be offered a sacrificed pig and some gold money. If the spirits are malicious and do not wish to leave the woman, the shaman must invoke his spirit assistants to negotiate with them. Usually, the evil spirits will be defeated and eventually agree to not interfere with the woman's body (womb).

Where infertility is due to disrupted social relationships, the offending woman must admit to having said bad things and vow to stop misbehaving while *ua neeb kho* (the healing ritual) is performed by a shaman. The shaman assists in removing the bad words and after this is done *txoov kab yeeb* will give children to the woman.

If the shaman succeeds in assisting with fertility the parents must return after the child is born to honour him. Traditionally, they must offer a sum of money, a bunch of incense sticks, gold money and a live chicken. The shaman continues to perform rituals to protect the newborn infant until it is three years old. During this period, if the child is ill the shaman is the first person to perform rituals to cure him/her. This tradition is still practised in Australia.

Discussion: Infertility, gender and social control

Although infertility is a biological problem, the experience of infertility is often mediated by cultural practices from within a society. Although infertility can occur in both male and female bodies, cross-cultural research reveals that women generally bear the burden (Inhorn, 1994a, 1994b, 2003; Neff, 1994; Kielmann, 1998; Prapreen et al, 2000; Riessman, 2000; Matsubayashi et al, 2001; Guntupalli and Chenchelgudem, 2004; Mariano, 2004; Johnson-Hanks, 2006; Rashid, 2007; Hollos and Larsen, 2008). Because of the expectation that all women are fertile, failure to conceive is usually seen as a problem inherent in a woman's body. For example, McGilvray (1982, p62) points out that among Tamil in Batticaloa, Sri Lanka, infertility is primarily seen as a woman's problem, although supernatural agency may be thought to contribute to her childlessness. The fertility of the husband is rarely questioned, and a woman without children is seen as both 'personally unfulfilled and ritually inauspicious'. Similarly, Neff (1994, p477) points out that for the Nayars in South India, the woman is most often blamed for her childlessness. The unfortunate woman will be an 'inauspicious guest at weddings and sacred rituals' and she is 'cut off from networks of sustenance and support'. Women in the Cameroon, according to Johnson-Hanks (2006, p249), 'are routinely brought to ritual specialists for infertility treatments' without any regard for their own desires or wishes. That infertility is a 'woman's problem' is clearly evident in Inhorn's study (1994a, p3) among poor urban Egyptian women. Women are blamed for their failure to reproduce, and they bear the burden of overcoming it through 'a therapeutic quest that is sometimes traumatic and often unfruitful'.

As I have illustrated, it is Hmong women, not men, who are blamed for being infertile, either because of the malfunctioning of the woman's body, her own bad behaviour, or her supernatural misfortune. Her 'afflicted body' provides legitimate grounds for the husband to bring in a second wife to bear children for him, and the childless woman has no say in this matter. Indeed, she has proven to him and his family that she may endanger the continuity of the family line, which in turn threatens the reproduction of Hmong society (Liamputtong and Spitzer, 2007).

Indeed, male infertility is rarely acknowledged in many societies, and perhaps particularly in patrilineal ones (Inhorn, 2003; Donkor and Sandall, 2007; Rashid, 2007). As Greer (1984, p34) argues: 'The reason why male sterility is seldom recognized is precisely because the recognition would strike at the heart of morality: all order and coherence would be put in jeopardy.' Similarly, men are thought to govern the Hmong world. Only men can perform important tasks and only men can carry on the clan name. The notion that men are infertile and therefore are unable to have their own sons (in particular) to continue their duties is seen as unacceptable, even immoral. Similarly, Inhorn (1994b) and Inhorn and van Balen (2002) suggest that infertility challenges men's procreative power. Women, through their 'faulty bodies' are nearly always blamed for not being able to bring men's children into the world. Therefore, male infertility is seldom recognized in Hmong culture, even though the women believe that the male can contribute to their barrenness as well.[4]

In other words, infertility in Hmong society is seen as the woman's problem. Infertility is an affliction of her body, her identity and on the 'local moral worlds' in which she lives (Kleinman, 1992; Inhorn, 1994b; Rashid, 2007). That her 'infertile body' in some ways reflects the patrilineal, patriarchal 'social body' is clear in the ways in which blame is assigned and control on the woman's body exercised (Douglas, 1970). At both the individual and societal levels, being fertile is a crucial part of Hmong life, for fertility ensures the continuation of the clan, lineage group and family – the three important pillars of Hmong society (Yang, 1992). Infertility, therefore, has profound implications for Hmong societal life. Women are seen as bearers of children, who must have as many children as they can. Failure to do so results in suffering both in this life and the next. For this reason, infertile Hmong women will try every ethnomedical cure within their reach to ensure their fertility.

At the time that I undertook my research, I found a strong maintenance of cultural norms among Hmong women and families despite living in a Western society. Although we have seen some social changes occurring in the Hmong community, most of the cultural beliefs and practices that I have discussed in this chapter continue to have an influence on Hmong women and their reproduction. As with the Hmong in the US (Thao, 1984; Fadiman, 1997; Vang et al, 2002; Cha, 2000, 2003; Culhane-Pera et al, 2004), the majority of Hmong in Australia continue to seek help from their traditional healers despite the availability of care within the Australian healthcare system. This is most obvious when the Hmong are confronted with severe illnesses and health-related issues that are seen to be closely related to the Hmong cosmos (such as in the case of infertility). But like their counterparts in the US, some Hmong who have more education and have lived longer in Australia have started to rely more on biomedical knowledge to deal with health and illness. And due to some difficulties in following their traditional practices, including the restriction of using animals in traditional healing rituals, and noises associated with the healing process, some Hmong have modified their traditions. I contend that accultura-tion among the Hmong in Australia has occurred, but it has been a slow process. How long the Hmong can maintain their cultural beliefs and practices remains to be seen. But for the time being, traditional beliefs and practices continue to have impact on the way Hmong women and family deal with infertility, as I have outlined in this chapter.

Acknowledgements

I am indebted to Hmong women who participated in this study and to Blia Ly, my bicultural research assistant who assisted me in the interviews and contributed knowledge about Hmong culture.

Notes

1   There was only one woman who became infertile after she had given birth to her first son in a refugee camp in Thailand.
2   Xoua Thao (1984) refers to this as 'Mandate of Life' in his writing of Hmong world views (see also Symonds, 1996, 2004).
3   One shaman with whom I spoke indicated that if a woman does not bring the paper with her even a shamanic healing process will not help her since it is she who does not want children and that is why she does not bring her paper to earth.
4   Although women are thought to be primarily responsible for infertility, men are occa-sionally mentioned as contributing to infertility if they have no testicles or uneven sized testicles. Some women explain that because of the uneven sized testicles, some men are not able to provide adequate seed for the woman. Because of this, these men cannot give children to their wives.

References

Bharadwaj, A. (2003) 'Why adoption is not an option in India: The visibility of infertil-ity, the secrecy of donor insemination, and other cultural complexities', *Social Science & Medicine*, vol 56, pp1867–1880
Cha, D. (2000) 'Hmong American concepts of health, healing, and illness and their ex-perience with conventional medicine', PhD thesis, Department of Anthropology, University of Colorado, Boulder, CO
Cha, D. (2003) *Hmong American Concepts of Health, Healing and Conventional Medicine*, Routledge, New York
Chapman, R. R. (2007) '"*Chikotsa*" – Secrets, silence, and hiding: Social risk and repro-ductive vulnerability in central Mozambique', *Medical Anthropology Quarterly*, vol 20, no 4, pp487–515
Chindasri, N. (1976) *The Religion of the Hmong Njua*, The Siam Society, Bangkok
Chindasri, N. (1992) 'Hmong shamanism', in J. McKinnon and W. Bhruksasri (eds) *Highlanders of Thailand*, Oxford University Press, Kuala Lumpur
Cooper, R., Tapp, N., Lee, G. Y. and Schworer-Kohl, G. (1995) *The Hmong*, ArtAsia Press, Bangkok

Culhane-Pera, K. A., Cha, D. and Kunstadter, P. (2004) 'Hmong in Laos and the United States', in C. R. Ember and M. Ember (eds) *Encyclopedia of Medical Anthropology: Health and Illness in the World's Cultures*, Kluwer Academic/Plenum, New York

Donkor, E. S. and Sandal, J. (2007) 'The impact of perceived stigma and mediating social factors on infertility-related stress among women seeking infertility treatment in Southern Ghana', *Social Science & Medicine*, vol 65, no 8, pp 1683–1694

Douglas, M. (1970) *Natural Symbols*, Pantheon Books, New York

Fadiman, A. (1997) *The Spirit Catches You and You Fall Down: A Hmong Child, her American Doctors, and the Collision of Two Cultures*, Farrar, Straus and Giroux, New York

Greer, G. (1984) *Sex and Destiny: The Politics of Human Fertility*, Secker and Warburg, London

Guntupalli, A. M. and Chenchelgudem, P. (2004) 'Perceptions, causes and consequences of infertility among the Chenchu tribe of India', *Journal of Reproductive and Infant Psychology*, vol 22, no 4, pp249–259

Gyekye, K. (1996) *African Cultural Values*, Sankofa Publishing Company, Accra, Ghana

Hmong National Development Links Inc. (2001) *HND Links: A Quarterly Newsletter*, Washington, DC

Hollos, M. and Larsen, U. (2008) 'Motherhood in sub-Saharan Africa: The social consequences of infertility in an urban population in northern Tanzania', *Culture, Health & Sexuality*, vol 10, no 2, pp159–173

Inhorn, M. C. (1994a) *Quest for Conception: Gender, Infertility, and Egyptian Medical Traditions*, University of Pennsylvania Press, Philadelphia, PA

Inhorn, M. C. (1994b) 'Interpreting infertility: Medical anthropological perspectives', *Social Science & Medicine*, vol 39, pp459–461

Inhorn, M. C. (1996) *Infertility and Patriarchy: The Cultural Politics of Gender and Family Life in Egypt*, University of Pennsylvania Press, Philadelphia, PA

Inhorn, M. C. (2003) '"The worms are weak": Male infertility and patriarchal paradoxes in Egypt', *Men and Masculinities*, vol 5, no 3, pp236–256

Inhorn, M. C. and van Balen, F. (eds) (2002) *Infertility around the Globe: New Thinking on Childlessness, Gender and Reproductive Technologies*, University of California Press, Berkeley, CA

Johnson-Hanks, J. (2006) *Uncertain Honor: Modern Motherhood in an African Crisis*, University of Chicago Press, Chicago, IL

Kielmann, K. (1998) 'Barren ground: Contesting identities of infertile women in Pemba, Tanzania', in M. Lock and P. A. Kaufert (eds) *Pragmatic Women and Body Politics*, Cambridge University Press, Cambridge

Kimani, V. and Olenja, J. (2001) 'Infertility: Cultural dimensions and impact on women in selected communities in Kenya', *The African Anthropologist*, vol 8, no 2, pp200–214

Kleinman, A. (1992) 'Local worlds of suffering: An interpersonal focus for ethnographies of illness experience', *Qualitative Health Research*, vol 2, pp127–134

Koster-Oyekan, W. (1999) 'Infertility among Yoruba women: Perceptions on causes, treatments and consequences', *Journal of Reproductive Health*, vol 3, no 1, pp13–26

Lee, G. Y. (2004) 'Culture and settlement: The present situation of the Hmong in Australia', in N. Tapp and G. Y. Lee (eds) *The Hmong of Australia*, Pandanas, Canberra

Liamputtong Rice, P. (1995) '"Pog laus, tsis coj khaub ncaws lawm": The meaning of menopause in Hmong women', *Journal of Reproductive and Infant Psychology*, vol 13, pp79–92

Liamputtong Rice, P. (1997) 'Giving birth in a new home: Childbirth traditions and the experience of motherhood among Hmong women from Laos', *Asian Studies Review*, vol 20, pp133–148

Liamputtong Rice, P. (2000) *Hmong Women and Reproduction*, Bergin & Garvey, Westport, CT

Liamputtong, P. (2004) 'Being a woman: The social construction of menstruation among Hmong women in Australia', in N. Tapp and G. Y. Lee (ed) *The Hmong of Australia*, Pandanas, Canberra

Liamputtong, P. (2007) 'Baby, souls, name and health: Customs for a newborn infant in Hmong society', in P. Liamputtong (ed) *Childrearing and Infant Care Issues: A Cross-Cultural Perspective*, Nova Science Publishers, New York

Liamputtong, P. and Spitzer, D. L. (2007) 'Double identities: The lived experience of motherhood among Hmong immigrant women in Australia', in P. Liamputtong (ed) *Reproduction, Childbearing and Motherhood: A Cross-cultural Perspective*, Nova Science Publishers, New York

Lock, M. and Scheper-Hughes, N. (1990) 'A critical-interpretive approach in medical anthropology: Rituals and routines of discipline and dissent', in T. M. Johnson and C. F. Sargent (eds) *Medical Anthropology: A Handbook of Theory and Method*, Greenwood Press, New York

Mariano, E. C. (2004) 'Involuntary childlessness among the Shangana (Mozambique)', *Journal of Reproductive and Infant Psychology*, vol 22, no 4, pp261–269

Matsubayashi, H., Hosaka, T., Izumi, S., Suzuki, T. and Makino, T. (2001) 'Emotional distress of infertile women in Japan', *Human Reproduction*, vol 16, no 5, pp966–969

McGilvray, D. B. (1982) 'Sexual power and fertility in Sri Lanka: Batticaloa Tamils and Moors', in C. P. MacCormack (ed) *Ethnography of Fertility and Birth*, Academic Press, London

Neff, D. L. (1994) 'The social construction of infertility: The case of the matrilineal Nayars in South India', *Social Science & Medicine*, vol 39, pp475–485

Okonofua, F. E., Harris, D., Odebiyi, A., Kane, T. and Snow, R. C. (1997) 'The social meaning of infertility in Southwest Nigeria', *Health Transition Review*, vol 7, pp205–220

Papreen, N., Sharma, A., Sabin, K., Begum, L. Ahsan, S. K. and Baqui, A. H. (2000) 'Living with infertility: Experiences among urban slum populations in Bangladesh', *Reproductive Health Matters*, vol 8, no 15, pp33–44

Quincy, K. (1995) *Hmong: History of a People*, Eastern Washington University Press, Cheney, WA

Rashid, S. F. (2007) '"Kal dristi", stolen babies and "blocked uteruses": Poverty and infertility anxieties among married adolescent women living in a slum in Dhaka, Bangladesh', *Anthropology & Medicine*, vol 14, no 2, pp153–166

Riessman, C. K. (2000) 'Stigma and everyday resistance practices: Childless women in south India', *Gender & Society*, vol 14, no 1, pp111–135

Symonds, P. V. (1991) 'Cosmology and the cycle of life: Hmong views of birth, death and gender in a mountain village in northern Thailand', PhD dissertation, Department of Anthropology, Brown University, Providence, RI

Symonds, P. V. (1996) 'Journey to the land of light: Birth among Hmong women', in P. Liamputtong Rice and L. Manderson (eds) *Maternity and Reproductive Health in Asian Societies*, Harwood Academic Publishers, Amsterdam

Symonds, P. V. (2004) *Gender and the Cycle of Life: Calling in the Soul in a Hmong Village*, University of Washington Press, Seattle, WA

Tapp, N. (2004) 'Hmong diaspora in Australia', in N. Tapp and G. Y. Lee (eds) *The Hmong of Australia*, Pandanas, Canberra

Tapp, N. and Lee, G. Y. (eds) (2004) *The Hmong of Australia*, Pandanas, Canberra

Thao, X. (1984) 'Southeast Asian refugees of Rhode Island: The Hmong perception of illness', *Rhode Island Medical Journal*, vol 67, pp323–330

Vang, C., Ichinose, T. and Murrieta, S. (2002) *Hmong Healthcare Practices in Orange County: Results from the Hmong Health Survey*, Public Health Services, Santa Ana, CA

Yang, D. (1992) 'The Hmong: Enduring traditions', in J. Lewis (ed) *Minority Cultures of Laos: Kammu, Lua', Lahu, Hmong, and Iu-Mien*, Southeast Asia Community Resource Centre, Rancho Cordova, CA

10

Experiences from a Constitutional State: Ireland's Problematic Embryo

Noëlle Cotter

Introduction

The use of new reproductive technologies (NRTs) by infertile couples in Ireland is shaped by particular socio-political meanings of procreation, which are bound up with pronatalist and pro-life discourses. Recent debates about the regulation of NRTs in Ireland have raised questions about the moral status of the embryo. These debates, although similar to those that have taken place in the UK (Mulkay, 1997; Parry, 2003), are distinctive, given the constitutional protection of the 'unborn' in Ireland. This Irish constitutional protection has created ambiguity around those embryos created extra-corporally, as in the case of NRTs. As a result, several stakeholder discourses have arisen in Ireland both around the in vitro fertilization (IVF) embryo itself and the potential for regulation of its creation, storage, usage and destruction, and these new discursive positions are the focus of this chapter.

The chapter therefore presents a discussion of the 'embryo question' in the Irish context, locating debates about the embryo created through technological means within the socio-historic specificity of the Irish case. Drawing on original research and secondary analysis of survey data and official documents, it considers perceptions of the embryo among four stakeholder groups: 'expert'; 'religious'; 'lay'; and 'patient' perspectives. The chapter begins with a brief discussion of the current Irish cultural context, particularly as it pertains to procreation, before turning to the 'embryo question'. It then discusses the four stakeholder positions in turn, before summarizing the discussion.

The Irish context

Ireland has a complex history which until recently was dominated by its relationship with the UK. European Union membership provided Ireland with a forum to assert its independence from the UK and has been fundamental to the nation's economic and social modernization. EU membership and foreign direct investment facilitated Ireland's economic growth engendering considerable social change. As a traditionally Catholic country, this era of modernization has led to conflict between competing worldviews within Ireland's social landscape, and a microcosm of this diversity is evident in issues surrounding human reproduction (McDonnell and Allison, 2006).

As an institution, the Catholic Church has lost much ground since the 1980s, with the recent economic boom and high profile clerical scandals possibly having contributed to declining support for the Church. Despite this decline, Catholic values and norms are embedded in Irish society and culture, and are reproduced through Irish institutions (Inglis, 1998). Without conscious adherence to the institution of the Church, Catholic mores are embedded in the Irish psyche and are potentially evident in Ireland's treatment of social issues which are of traditional concern to Catholicism; in particular issues pertaining to marriage, sexuality and procreation.[1] This influence is particularly slow to dissipate where Catholicism is enshrined in law.[2] For example, although citizens are entitled to abortion information, to travel for abortions and to undertake abortions where the life of the woman is at risk, abortion is still illegal and unconstitutional in Ireland, despite referenda in 1983, 1992 and 2002 (Oaks, 2003). In the 1983 referendum on abortion, the following text was inserted into the constitution and remains today as Article 40.3.3:

> *The State acknowledges the right to life of the unborn and, with due regard to the equal right to life of the mother, guarantees in its laws to respect, and, as far as practicable, by its laws to defend and vindicate that right.*

The constitutional protection of the 'unborn' has implications for IVF/ICSI (intracytoplasmic sperm injection) and the use of embryos created by these techniques beyond the original intent of this insertion – which was to prevent abortion on Irish soil. The way in which NRTs have been accepted and negotiated in the Irish context is of interest, given this historical position on abortion and persistent pro-life discourses. In particular, the implications of technologies such as IVF, for the creation, storage and disposal of embryos, has recently sparked debate in Ireland and it is these debates that are the focus of this chapter.

The Irish regulatory vacuum: Embryos as boundary objects

Ireland has no official body or legislation for the regulation and licensing of clinics offering IVF or other reproductive technologies, despite their presence in the country since the late 1980s. Thus far, physicians practising in this area have been provided with guidelines by the Irish Medical Council. Without specific regulation, technically there is *carte blanche*, with physicians governed only by industry self-regulation. The Irish Medical Council specifies that physicians are not permitted to destroy 'fertilized ovum' – in case these ovum incur constitutional protection as 'unborn' (see Irish Medical Council,1981–2004). The Irish Medical Council is therefore essentially treading water, waiting for clarification of embryo status. This situation means that at present, if deemed viable, embryos created through NRTs on Irish soil must be cryopreserved indefinitely or used by the patient.

To address this lack of regulation, in 2000 the Irish Minister for Health and Children established the Commission on Assisted Human Reproduction (CAHR). Similarly to the UK's Warnock Commission in 1982, CAHR's terms of reference stated they were to prepare a report on the possible approaches to the regulation of all aspects of assisted human reproduction and the social, ethical and legal factors to be taken into account in the determining of public policy in this area (CAHR, 2005). During this period of regulatory uncertainty, a number of stakeholders emerged, each of whom has a varying degree of investment in the definition of the embryo within Irish policy. For the purposes of this chapter, these interest groups are represented in the following way: the 'expert' stakeholders, comprising those who are invited by the state or official body to provide expert opinion on issues relating to reproductive technologies (in this case CAHR and the Irish Council for Bioethics), which includes scientists, physicians, academics, civil and public servants; the 'religious' stakeholder group, represented here via a discussion of the Irish Catholic bishops' response to CAHR recommendations; 'lay' stakeholders, that is, members of the public who do not have direct experience of this issue but who potentially hold power based on their right to vote should a referendum occur; and 'patient' stakeholders, who are users of IVF, and who would be directly impacted by regulation and legislation but whose voice has been almost entirely silent in Ireland to date.[3] Despite discursive momentum around this issue, there has been as yet, no political move towards addressing the embryo question, or adopting CAHR's recommendations.[4] This chapter will illustrate how these positions demonstrate the contested nature of embryos, something which is not unique to Ireland.

Setting aside national regulations, there are six options broadly considered available to couples with 'excess' embryos. These are: use in own treatment; placing in the uterus when implantation is impossible for the purposes of disposal; donation to another person(s); donation to infertility research; donation to general research; destruction (Nachtigall et al, 2005). When national regulations are taken into account, there are fewer options. Couples may not have multiple options available to them, and may also be governed by cryopreservation time limits. For example, Bangsbøll et al (2004) describe how the Danish 24-month cryopreservation time limit contributes to the destruction of 2500 embryos each year. Many couples, upon becoming parents, would not be able to consider having another child so quickly. The option of birth spacing is removed in this example. In the UK, the cryopreservation limit is five years (with possible extension to ten years) in which time couples can decide to use their embryos in their own treatment, allow them to perish, donate them to research or donate them to another couple (Parry, 2003). This allows couples who may have completed treatment some sense of 'closure', whereas for others it may be experienced as an arbitrary regulatory imposition (de Lacey, 2007). As described earlier, in the Irish context couples have only two options: use viable embryos in treatment or cryopreserve indefinitely.

Ireland is not alone in experiencing the embryo as problematic. The birth of Louise Brown in the UK in 1978 created a legislative vacuum, and associated technologies created a moral panic (Franklin, 1993). UK parliamentary debates about embryos grew out of the abortion debates preceding the 1967 Abortion Act, resulting in the Human Fertilisation and Embryology (HFE) Act 1991. These debates provided a new platform for a dormant anti-abortion lobby to agitate on (Mulkay, 1997). Similarly in Ireland, consideration of embryos as contentious entities has grown out of the abortion debates. The difference between the UK and Irish experience in these debates is principally that in the UK, the pro-abortion lobby was successful, and in Ireland it was unsuccessful. A country's political position on abortion is therefore of considerable significance when considering the embryo question.

In 2001, the HFE Act was amended to facilitate the development of stem cell research in the UK, and at this time the use of similar rhetoric and strategies, which attempt to claim various statuses and moral worth for the embryo, were deployed once again (Parry, 2003). Parry (2003) identifies how 'boundary' work facilitated the opposing sides in the debates' common focus on a 'return to the embryo question' in the context of stem cell technologies. The Irish situation is similar to the UK context, in that an embryo debate generates a number of competing claims over the embryo, and is entwined with debates about abortion. However, the Irish situation is unique in that Ireland is a predominantly Catholic country, abortion is unconstitutional and embryo research does not take place in Ireland.

Other authors have used the concept of the 'boundary' in discussing human embryos, in order to demonstrate how they represent different entities for different groups (Williams et al, 2008). Williams et al (2008) describe how embryos as 'boundary objects' facilitate the interface between the related social worlds of embryonic stem cell laboratories and pre-implantation genetic diagnosis laboratories. Boundary objects are entities that are sufficiently flexible to retain commonality across worldviews but are perceived differently from each perspective (Williams et al, 2008). Embryos are 'constitutively promissory' (Thompson, 2005); their value across all boundaries of understanding is in their potential. Thus embryos can be perceived as potential humans (along a continuum of understood potentiality), but also as providing potential disease cures. Within the Irish context, embryos as 'boundary objects' are sufficiently flexible to exist in different social worlds (expert, religious, lay and patient), but their interpretation varies and is influenced by cultural Catholicism.

This chapter argues that embryo perceptions are dependent on the stakeholder, the perspective and the context, and in Ireland are further complicated by Ireland's constitution and traditional adherence to Catholicism. The idea of the embryo as a 'boundary object' in the Irish context will therefore be considered. This chapter will discuss 'expert', 'religious', 'lay' and 'patient' views of the embryo, demonstrating competing Irish discourses around this issue. Although these discourses will be discussed as separate entities in this chapter, this is not to suggest there is cohesion or mutual exclusivity within these views, nor that these are the only discourses.[5] The expert, religious and lay views will be briefly considered by drawing on secondary survey data and published reports. The patient perspective will be discussed through an analysis of primary interview data carried out by the author.

The 'expert' perspective

In 2005, CAHR published its report recommending that a regulatory body should be established to oversee the use of reproductive technologies, such as IVF, in Ireland. The report included 40 recommendations, none of which have been implemented. The most significant of these recommendations for this debate stated:

> *The embryo formed by IVF should not attract legal protection until placed in the human body, at which stage it should attract the same level of protection as the embryo formed in vivo.*

This recommendation, if adopted, would have implications for the introduction of embryo research and disposal in Ireland. The CAHR members agreed that this was a problematic recommendation which could not be easily adopted in the light of the constitutional protection of the 'unborn' in Irish law. This protection would need to be clarified by referendum[6] or Supreme Court decision, if CAHR's recommendation were to be adopted. CAHR suggested that it is not clear whether protection of the 'unborn' applies from the point of fertilization or some point thereafter, and this would therefore need to be defined. According to CAHR's recommendations, this means categorizing the 'unborn' as the embryo in vivo (in the body), and not the embryo in vitro, rendering embryos created outside of the body free from constitutional jurisdiction. The majority of members of this 'expert' group therefore regarded the embryo in vitro (as in the case of IVF) as lying outside this constitutional definition of 'unborn'.

More recently, the Irish Council for Bioethics (2008) produced a report on stem cell research. In this, Council members supported the practice of stem cell research in Ireland and encouraged the development of legislation to govern it. Although there is some dissent, albeit marginal, the expert discourse of the embryo in Ireland constructs the embryo as unproblematic and morally neutral whilst outside of the body, only attaining status of 'unborn' when in vivo.

The responses of CAHR and the Irish Council for Bioethics members to the embryo question illustrate the discourse of 'science as progress' characterizing these groups' position on the embryo. Their professional backgrounds in science and academia may contribute towards their construction of research and its endeavours as inherently beneficial.

The 'religious' stakeholders: Institutional Catholicism

In response to issues pertaining to reproductive technologies, the Catholic Church has globally had an active voice. In order to demonstrate this voice in the Irish context, this perspective is represented here by considering the response of the Irish bishops to CAHR's recommendations (Irish Catholic Bishops Conference, 2006), which contains within it their responses to CAHR's 40 recommendations. In this report, the bishops rejected the majority of CAHR's recommendations for a number of reasons, which can be summarized in the interpretation of the following text from the Vatican document *Donum Vitae* (Instruction on Respect for Human Life in its Origins and on the Dignity of Procreation):

> *The child has the right to be conceived, carried in the womb, brought into the world and brought up within marriage, it is through the secure and recognised relationship to his own parents that the child can discover his own identity and achieve his own proper human development* (cited in Irish Catholic Bishops' Conference, 2006, p9)

Specifically, the Irish Catholic bishops could not support CAHR's perspective of the embryo whilst in vitro:

> *Once fertilisation has been completed a new human being exists, and this brings with it an obligation of respect. It is clearly in the interests of justice and the common good that this obligation should be reflected in civil law.* (2006, p6)

The bishops were clear that the embryo should not be differently configured by location, and that human life begins at the moment of fertilization, affording it constitutional protection as 'unborn', mirroring the Vatican's stance and clearly reiterated by Pope Benedict XVI in addressing the general assembly of the Pontifical Academy for Life in February 2006.[7]

The 'lay' perspective

Both CAHR and the Irish Council for Bioethics commissioned surveys to gauge public opinion in relation to the embryo issue, and these data have been analysed and presented here to represent the 'lay' or public response. Public opinion has therefore only been evident in Ireland to date when the 'expert' groups seek lay opinions for official documentation.

In 2005, the Irish Council for Bioethics released the results of a nationally representative survey gauging public opinion across a wide range of contentious medical practices (Irish Council for Bioethics, 2005). One of these practices was assisted human reproduction. This survey was carried out between August and September 2005 with 500 adults over age 18 who were interviewed face-to-face using a structured questionnaire. The sample was quota controlled in terms of gender within age within region, and socio-economic controls were incorporated. The findings revealed that 57 per cent of the Irish public believed life began at conception, 19 per cent believed life begins at implantation, 13 per cent believed life began three months into a pregnancy and 4 per cent believed life began at birth. These results reveal that the majority of the Irish public believe life begins from the point of fertilization, and is not based on being in vivo. However, the caveat to this inference is that the Irish public are not necessarily well versed in the practice of IVF, and the boundaries of 'life' may shift in response to a full media debate on the topic prior to a referendum, should it occur, to clarify Article 40.3.3. This caveat also applies to the public attitudes survey commissioned and utilized by CAHR. Its survey revealed that 45 per cent of the Irish public were opposed to the production of surplus embryos in the course of IVF treatment, and 48 per cent were opposed to the freezing of surplus embryos. Though limited, these data indicate there is not widespread support for some fundamental elements of IVF practice.

The Irish Council for Bioethics also attempted to gauge public opinion for their stem cell research report published in 2008 (Irish Council for Bioethics, 2008). The Council invited public submissions in the form of a questionnaire. This meant that respondents were self-selecting and therefore is not necessarily representative of the Irish public; however, responses followed a similar pattern to the survey published earlier by the same body (Irish Council for Bioethics, 2005). Of 2200 questionnaire submissions, 69 per cent believed the embryo acquires full moral status at fertilization, only 9.7 per cent believing full moral status was acquired when an embryo was implanted in the womb. When asked about using IVF-derived embryos for stem cell research which would lead to their destruction, 70.6 per cent did not believe that this practice should occur.

As in the UK case, Irish opposition to embryo production/research/destruction may have emerged from the remnants of the anti-abortion movement, but in combination with a cultural background of Catholicism, biotheology (Franklin, 1993) may intensify public concern regarding the embryo in Ireland. A large percentage of the Irish public may therefore see the embryo as a moral entity, and as a potential citizen attracting constitutional protection.

The 'patient' perspective

Of the 34 couples (see Note 3) who took part in the author's research to completion, almost all had the experience of embryo creation and cryopreservation. For these patients, the knowledge that fertilization had occurred in the laboratory was highly significant, surpassed only by the survival to implantation/cryopreservation of their IVF embryos. These couples had spent a long time attempting to get pregnant prior to beginning their IVF/ICSI cycles and were reliant on pregnancy tests to tell them, retrospectively, if pregnancy (fertilization and implantation) had occurred – which it rarely had. IVF/ICSI differs from conceiving 'naturally' and less high-technology attempts at conception (e.g. intrauterine insemination (IUI)), in that couples know within 24 hours after egg and semen harvesting if fertilization has occurred, and are guaranteed that embryos returned to the uterus and/or cryopreserved are 'promissory'. In other words, they contained the promise of a pregnancy and live birth outcome. Therefore, the following sections will discuss how patients perceived embryo creation as a milestone and how potentiality and place also proved significant, with similarities to already discussed discourses; providing further evidence of embryos as 'boundary objects'.

Embryo creation as a milestone

For couples undergoing IVF treatment, the point of embryo creation and transfer is a huge milestone. Sandelowski (1993) describes IVF treatment as a series of pregnant moments. As treatment progresses and couples get past each phase, hope builds and in the final stages of contact with the clinic, creation and transfer of embryos to the uterus is the penultimate milestone. IVF can be experienced as an obstacle course (Franklin, 1997) or as hurdles (Monach, 1993), meaning that each stage of treatment can be experienced as success or failure in the 'pregnancy trail' (Parry, 2006).

For couples with identified reproductive impairments who do not therefore have an opportunity to become pregnant in any given month, the moment of discovering they have embryos for implantation is highly significant as this may be their only chance at becoming pregnant. Sandelowski (1993) describes this as 'ambiguous conception'; there is a pregnancy continuum which is experienced prospectively by couples undergoing IVF/ICSI treatment. Thus, in these circumstances, embryos are valued outcomes of treatment; their creation is the potential precursor to a much desired pregnancy and they are frequently considered as potential, or actual, life. One couple in the study described how they were caught unawares by how strongly they felt about their embryos and how the creation of embryos was a significant achievement:

> Mary: ... *I hadn't ever considered what it would feel like to actually suddenly have an attachment to eight embryos. I mean we'd gone from nothing, from no children to eight potential children, who may have come to nothing as well. But for that moment in time and that day, there were eight. A combination of boys, girls, you know, light-haired, dark-haired, all those things were already decided, that just hadn't crossed my mind.*

> Paul: ... *you feel like a sense of achievement and if nothing else works and this whole process we've been through, we can say that we've got to that stage ... to date we've had five or six IUIs and nothing happened. And we don't know whether they actually got fertilized or if they didn't get fertilized ... it's a strange feeling that you think okay at least we've managed it – in a fridge somewhere [laughter]. You know, mixed our DNA which is a very strange thing...*

Mary and Paul's experience of trying to conceive and of infertility treatment has been a long process of consistent failure. The creation of embryos suddenly offered them hope; they were proven capable of achieving something that they had no previous knowledge of – his sperm could successfully fertilize her eggs. In this event, common understandings of kinship come to the fore, evident in this couples' interpretation of genetics. Their DNA had successfully mixed, offering them the potential of their own child, born of their own biologic make-up. This was a 'pregnancy moment', but the hurdle of implantation was yet to be surmounted.

Potentiality and place

For some of the couples undergoing IVF, the creation of embryos was not simply a creation of cells. It took on a greater significance. Although aware of the scientific perception of embryos, and having placed their confidence in science, some couples maintained a more spiritual or moral belief in life and its beginnings. Couples described their embryos as 'potential babies' or 'potential life' and in a small number of cases used the word 'babies' when describing their embryos. This discursive construction of embryos as human life has resonances with the position held by the Catholic Church. However, few participants discussed their embryos in relation to religious beliefs. Their views of embryos as life were bound up with potential that they had to (re)produce the couples as parents (Becker, 2000; Thompson, 2005).

There was a range of opinions expressed when couples were asked directly how they felt about their embryos after the point of implantation/cryopreservation. Fertilization was viewed as a milestone with great significance for all couples. The 'obstacle course' of IVF means that couples who achieve embryo creation feel great excitement at reaching this stage (Franklin, 1997). However, if at the point of fertilization embryos did not 'progress', these embryos were discounted; they were regarded as never having the potential for life; they had been 'naturally' selected out. These embryos were regarded as fertilized eggs that were never intended, almost predestined to not come into being as a potential life or, further along, as a baby. Parry (2006) describes this distinction in terms of embryos being disconnected from the 'pregnancy trail'. The couple cited below compare 'non-viable' embryos with what could occur any month without their knowledge, but make a distinction between these embryos and their frozen embryos:

> Jane: *I think it's probably like in a cycle if you're producing an egg every month and if they don't take you know …*

> Peter: *Yeah.*

> Jane: *Hum – I don't really think of them you know at all. Like I suppose we had our grief when it failed the first time but I never thought of the others that weren't implanted you know. Now, I feel terrible strongly about the ones that are frozen.*

> Peter: *Oh yeah, yeah.*

Compared to embryos that never were, couples felt strongly about both their viable embryos that were freshly implanted, as well as about the frozen embryos, which they planned to have implanted at a later date. Strong feelings about frozen embryos were further reinforced when treatment was successful. In these cases, frozen embryos were regarded by some respondents as the twin(s) of their existing child(ren) as they had been created at the same time. They would find it (hypothetically) difficult to destroy what they perceived as being a sibling to their child(ren). One woman, who had three successful ICSI pregnancies, described her desire to use her frozen embryos as needing to 'gather my chicks', and, through tears, described how she would always mourn the embryos that either did not implant or did not survive thawing.

When embryos were thawed, survived and were placed in the uterus, they were regarded in the same way by couples as embryos placed in the uterus during a fresh cycle. When embryos did not survive the thawing process, although now technically in the same category as embryos that did not survive long after fertilization, couples felt much more deeply about them. One woman described the cryopreservation process as 'suspending life' and this could explain the differentiation between a pre-implantation fresh non-surviving embryo and a pre-implantation frozen non-surviving embryo; hope had been invested in these frozen embryos over a longer period of time, facilitating a new understanding of them. The original fresh cycle non-surviving embryos were life that was never meant to be, but embryos that survived to the cryopreserved state were regarded by some couples as being the same potential life as freshly implanted embryos. The fact that they did not survive the thawing process implied that the process was too rigorous, not that the embryo was not meant to 'live'. It was not common among the couples, but a small number of respondents felt the need to mourn embryos that did not survive the thawing process, or that did not implant. One couple shared their need to have a private ceremony at home where they lit candles and named their embryos, marking their time in existence before going to a 'better place':

> Lisa: *Well it is over but I suppose you'll just, there'll always be kind of you know the six that never were you know, em.*

> Phil: *Well they were at some level, it depends what you believe I suppose ... what I would believe a little bit is they came in to live in whatever experience and they've experienced and that's all been helpful for whatever they were doing and whatever we were doing so you know they didn't survive so that was fine, the candles is just a closure thing.*

Confirmed pregnancy as 'unborn'

It is necessary to consider the small number of respondents in the sample who did not regard their embryos as anything more than a cluster of cells until a positive pregnancy test. There were five couples in the sample who were in agreement that their embryos were not considered life until a definite pregnancy was confirmed.

Pat: *No, I suppose I would be thinking that you know, we had fertilization and it divided into two and it divided into four. Well you know that's a little bit kind of a little bit early to kind of portray anything other than, really it's still only an egg and a sperm, really it's only still at that stage. And it would have been the next stage before I got any feelings I think …*

Ann: *And I think I suppose that some people would say that technically the day they implant them, technically you could consider yourself pregnant but I never would have… No, not until you had done the two weeks to get the test or whatever I wouldn't have, no. And I've always known before the two weeks which I said before so I never had any sense of 'gosh I feel pregnant' or kind of feel I might be or whatever, never had any sense of that. So I don't feel they ever implanted even for ten days or whatever it was and it was in some ways to me it's kind of irrelevant, no, if I had got a positive pregnancy test then I would feel you were pregnant at some point, but no.*

A small number of male partners disagreed with their female partners in this regard; they focused on the scientific explanation of the embryo as a fertilized egg and did not perceive embryos as life.

Patients' perceptions of embryos were clearly influenced by the context of emotions associated with undergoing IVF/ICSI; perceiving the treatment as a series of pregnant moments (Franklin, 1993; Sandelowski, 1993). Perceptions of embryos often fused both the scientific and the spiritual. Couples had placed their confidence in science and understood the scientific perception of embryos as a means to an end; as pre-life. But couples also had an understanding that it was not science or its endeavours that had created these embryos – these embryos were the result of the combination of their gametes and their DNA, and so held greater value. The daily, ordinary 'miracle' of such occurrences is more abstract. Biomedicine is a facilitator for what can occur, and usually does, in more 'natural' circumstances – the miracle of life, as it is perceived, still occurs with IVF, but it is in vitro, instead of in vivo. Biomedicine can bring gametes together, and in the case of ICSI, fusion can be forced, but the success of this fusion is outside of biomedicine's control. It is more than the sum of its parts. Hence, embryos were value-laden for some couples in this sample as potential children – their kin. As Edwards (1993) demonstrates, we are all kinship experts by virtue of our common understanding of the cultural importance of blood links, even if we are not scientific experts.

These interviews demonstrate the highly nuanced nature of patients' perceptions of embryos. Couples had very specific attitudes towards their embryos, based on their location and survival beyond initial fertilization. If a continuum can be placed on the data, it would appear that the initial discovery that embryos had been created is very emotional, but the survival of these embryos to transfer or cryopreservation is more significant. Embryos that disintegrated soon after formation had considerably less significance. This fusion was not destined to become a potential child. It must not be forgotten that this sample also included a small minority of people who did not consider embryos significant until a positive pregnancy test. This is not to suggest that scientific rationality was entirely pervasive among these respondents; embryo creation and fragmentation was still highly significant and emotionally charged. Therefore for patients in this study, initial embryo creation gave hope and was a major milestone; however it was implantation and cryopreservation that imbued the embryo with potential.

Despite the specificity of culture, options and timing of research in this case, some of the findings presented here are similar to those from other studies. Internationally, Söderström-Anttila et al (2001), McMahon et al (2002), Nachtigall et al (2005) and Parry (2006) also found the view of embryos as potential children among couples making decisions about 'surplus' embryos. However, research in this area also reports perceptions of embryos as 'biologic material' (Nachtigall et al, 2005).

Conclusion

This chapter has discussed the Irish cultural context in which the embryo is potentially a contested entity. At the core of this milieu are Ireland's traditional roots in Catholicism, which has heavily influenced, and still influences, the Irish worldview. Although in decline, Catholicism will be slow to disappear considering its embedded position in Irish institutions and constitution. Within Ireland's own emerging discourses about the embryo, this chapter has focused on four principal stakeholder groups; the 'expert', 'religious', 'lay' and 'patient', each with a stake in the embryo question and each (re)creating the embryo as a boundary object in the Irish context.

The 'expert' discourse is constituted of a legalistic, scientific and regulatory response to the problematic embryo. The perception here is that the wording of the constitution is delaying 'progress' and without directly tackling the problem there is a regulatory vacuum, which is perceived as somewhat dangerous by this group of stakeholders. The religious discourse, on the other hand, clearly positions all embryos as human life, affording them civil and moral protection. For the religious stakeholders, therefore, embryos have fundamental potential, regardless of place. The 'lay' discourse suggests that the Irish public may adhere to the definition of life as beginning at the point of fertilization, which may challenge the practice of IVF/ICSI in the state, and also the availability of certain forms of contraception, should legal definitions of embryos be delineated (e.g. the morning-after pill).

IVF patients are intimately involved in embryo creation in a way that the other stakeholder groups are not, since for this group (certain types of) embryos are unique in the reproductive and parenting potential that they encapsulate. Within this sample there was a common theme regarding embryo creation as a milestone, but a divergence of views over when the potential for life begins, with some patients aligning themselves with a more scientific or 'expert' view, and some with a more 'religious'/spiritual or moral view. The distinction made between embryos that do not survive long after fertilization, and those that do not survive cryopreservation, can be understood in terms of suspending life.

Cryopreservation may not be interpreted as freezing at a point in time instead but – with more complexity – as suspending a potential life that is waiting to come into being. There is an inherent contradiction between the patient and the religious discourse, therefore, on the significance of freezing, since for the Church freezing embryos amounts to a lack of respect for human life, whereas for patients, cryopreservation meant that embryos were even more precious. More research is required to fully comprehend patient views in this respect, especially since for patients in this study, their embryos were still being used in treatment, and were not considered to be 'surplus'.

In this discussion of competing Irish discourses, perceptions of embryos among 'expert', 'religious', 'lay' and 'patient' groups have been explored, and operation of the embryo as a boundary object has proven insightful. Although there are similarities with findings from other international contexts, in particular those experienced in the UK in the past two decades, the Irish case is unusual given the constitutional protection afforded to the 'unborn' and a lack of definition in this regard.

If a referendum takes place, the word 'unborn' will need to be defined, and in addressing this, the Irish voter may be asked to decide again if they are willing to allow abortion on Irish soil. By clearly stating the word 'abortion' it may be possible to separate these two issues: abortion and the use of embryos. However, in defining the word 'unborn' it would be necessary to decide if, as a nation, Irish citizens are happy for a regulating authority or individual to make decisions about embryos – their destruction, donation or experimentation. Alternatively, the decision, as a nation, to protect all embryos, regardless of place, will need to be made. Above all, it will be necessary to find the political will to tackle this issue.

Notes

1   For example, divorce was not legal in Ireland until the 1995 referendum (50.28 per cent voted in favour). The 1979 Health (Family Planning) Act allowed access to contraception via a physician's prescription and condoms were not allowed to be sold without prescription until 1985 and then only from named outlets. The sale of condoms was not deregulated until 1992. See Coakley (1996), Inglis (1998), Mahon et al (1998), Oaks (2003), for further details on divorce, contraception and abortion in Ireland.

2   The Catholic Church has traditionally been a 'power bloc' with a 'moral monopoly' in Irish society (Inglis, 1998). Ireland's constitution (Constitution of Ireland – Bunreacht na hÉireann, 1937) stated 'The State recognises the special position of the Holy Catholic Apostolic and Roman Church as the guardian of the Faith professed by the great majority of the citizens'. Until the deletion of this passage in 1972 there was a clear relationship between Church and State enshrined in the constitution. This influence is still present, albeit in more subtle tones (Gallagher, 1996).

3   Data discussed were garnered from a number of principal sources detailed in the text; official Irish government and public bodies' publications concerning assisted human reproduction, and the public attitude surveys and submissions utilized in these publications as well as the Irish Catholic Church's response. 'Patient' perspectives were

drawn from research on couples' experiences of infertility and its treatment in Ireland. This research is longitudinal; 34 couples were interviewed at four time points – before beginning an IVF/ICSI cycle, after embryo transfer and while waiting to do a pregnancy test / in the immediate aftermath of a cancelled cycle, three months later, and a further 12–18 months later. The embryo issue arose initially without prompting, but proved so pertinent that it became incorporated into the topic guide, proving to be a highly emotive topic to be dealt with delicately, if at all, at the risk of upsetting respondents. As with most qualitative research, data from these interviews are not representative; all respondents volunteered for this research and were recruited from one Irish infertility clinic. All couples were married at the time of the first interview – with the exception of one couple, but they married prior to starting a treatment cycle. All couples were of ethnic origin 'White'. Twenty-five couples were Irish by birth, seven couples had one Irish partner with the other originating in another country, and two couples where neither party was born in Ireland were also in this sample. Couples were not directly asked their religious affiliation and further demographic information is excluded in order to protect the anonymity of the participants.

4   A test case before the courts may eventually force the political hand; in 2007 a separated couple were before the Irish courts contesting the use of frozen embryos by the female where the husband had withdrawn consent. The court ruled in the husband's favour but the fate of these embryos is still undecided. *Roche* v. *Roche – High Court Judgment,* www.courts.ie/Judgments.nsf/09859e7a3f34669680256ef3004a27de/98ecc05ed4417bbc8025729600444e2a?OpenDocument, accessed 3 February 2008.

5   This is not to suggest that there are only four stakeholder groups, or that these are distinct. There is much fluidity between these, and there may be other stakeholder groups meriting consideration (e.g. the dormant pro-abortion lobby) but brevity must dictate.

6   A referendum – a national vote – is necessary for any change to the Irish Constitution.

7   Address of His Holiness Benedict XVI to the participants at the 12th General Assembly of the Pontifical Academy for Life and Congress on 'The Human Embryo in the Pre-Implantation Phase' (27 February 2006), www.vatican.va/holy_father/benedict_xvi/speeches/2006/february/documents/hf_ben-xvi_spe_20060227_embrione-umano_en.html, accessed 3 July 2008.

References

Bangsbøll, S., Pinborg, A., Yding Andersen, C. and Nyboe Andersen, A. (2004) 'Patients' attitudes towards donation of surplus cryopreserved embryos for treatment or research', *Human Reproduction*, vol 19, no 10, pp2415–2419

Becker, G. (2000) *The Elusive Embryo: How Women and Men Approach New Reproductive Technologies*, University of California Press, Berkeley, CA

CAHR (2005) 'Report of the Commission on Assisted Human Reproduction', www.dohc.ie/publications/cahr.html, accessed 3 February 2008

Coakley, J. (1996) 'Society and political culture', in J. Coakley and M. Gallagher (eds) *Politics in the Republic of Ireland*, PSAI Press, University of Limerick, Ireland

Constitution of Ireland – Bunreacht na hÉireann (1937), www.taoiseach.gov.ie/index.asp?docID=243, accessed 3 February 2008

de Lacey, S. (2007) 'Patients' attitudes to their embryos and their destiny: Social conditioning?', *Best Practice & Research Clinical Obstetrics and Gynaecology*, vol 21, no 1, pp101–112

Edwards, J. (1993) 'Explicit connections: Ethnographic enquiry in north-west England', in J. Edwards, S. Franklin, E. Hirsch and F. Price (eds) *Technologies of Procreation: Kinship in the Age of Assisted Conception*, Manchester University Press, Manchester

Franklin, S. (1993) 'Making representations: The parliamentary debate on the Human Fertilisation and Embryology Act', in J. Edwards, S. Franklin, E. Hirsch and F. Price (eds) *Technologies of Procreation: Kinship in the Age of Assisted Conception*, Manchester University Press, Manchester

Franklin, S. (1997) *Embodied Progress: A Cultural Account of Assisted Conception*, Routledge, London

Gallagher, M. (1996) 'The Constitution', in J. Coakley and M. Gallagher (eds) *Politics in the Republic of Ireland*, PSAI Press, University of Limerick, Ireland

Inglis, T. (1998) *Moral Monopoly: The Rise and Fall of the Catholic Church in Modern Ireland*, University College Dublin Press, Ireland

Irish Catholic Bishops Conference (2006) *Towards a Creative Response to Infertility. A Detailed Response of the Irish Catholic Bishops' Conference to the Report of the Commission on Assisted Human Reproduction*, Veritas, Dublin, www.catholiccommunications.ie/creativeresponse/towardsacreativeresponse.pdf, accessed 19 June 2008

Irish Council for Bioethics (2005) 'Bioethics Research', www.bioethics.ie/pdfs/129171-Bioethics%20Research.pdf, accessed 3 February 2008

Irish Council for Bioethics (2008) *Ethical, Scientific and Legal Issues Concerning Stem Cell Research: Opinion*, Irish Council for Bioethics, Dublin

Irish Medical Council (1981, 1984, 1989, 1994, 1998, 2004) *Constitution and Functions: A Guide to Ethical Conduct and Behaviour and to Fitness to Practice*, Irish Medical Council, Dublin

McDonnell, O. and Allison, J. (2006) 'From biopolitics to bioethics: Church, state, medicine and assisted reproductive technology in Ireland', *Sociology of Health and Illness*, vol 28, no 6, pp817–837

Mahon, E., Conlon, C. and Dillion, L. (1998) *Women and Crisis Pregnancy*, The Stationery Office, Government publications, Dublin

McMahon, C. A., Gibson, F. L., Leslie, G. I., Saunders, D. M., Porter, K. A. and Tennant, C. C. (2002) 'Embryo donation for medical research: Attitudes and concerns of potential donors', *Human Reproduction*, vol 18, no 4, pp871–877

Monach, J. H. (1993) *Childless: No Choice. The Experience of Involuntary Childlessness*, Routledge, London

Mulkay, M. (1997) *The Embryo Research Debate: Science and the Politics of Reproduction*, Cambridge University Press, Cambridge

Nachtigall, R. D., Becker, G., Friese, C., Butler, A. and MacDougall, K. (2005) 'Parents' conceptualization of their frozen embryos complicates the disposition decision', *Fertility and Sterility*, vol 84, no 2, pp431–434

Oaks, L. (2003) 'Antiabortion positions and young women's life plans in contemporary Ireland', *Social Science & Medicine*, vol 56, pp1973–1986

Parry, S. (2003) 'The politics of cloning: Mapping the rhetorical convergence of embryos and stem cells in parliamentary debates', *New Genetics and Society*, vol 22, no 2, pp145–168

Parry, S. (2006) '(Re)constructing embryos in stem cell research: Exploring the meaning of embryos for people involved in fertility treatments', *Social Science & Medicine*, vol 62, pp2349–2359

Sandelowski, M. (1993) *With Child in Mind: Studies of the Personal Encounter with Infertility*, University of Pennsylvania Press, Philadelphia, PA

Söderström-Anttila, V., Foudila, T., Ripatti, U. R. and Siegberg, R. (2001) 'Embryo-donation: Outcome and attitudes among embryo donors and recipients', *Human Reproduction*, vol 16, no 6, pp1120–1128

Thompson, C. (2005) *Making Parents: The Ontological Choreography of Reproductive Technologies*, MIT Press, Cambridge, MA

Williams, C., Wainwright, S. P., Ehrich, K. and Michael, M. (2008) 'Human embryos as boundary objects? Some reflections on the biomedical worlds of embryonic stem cells and pre-implantation genetic diagnosis', *New Genetics and Society*, vol 27, no 1, pp7–18

11

Marginalized, Invisible, and Unwanted: American Minority Struggles with Infertility and Assisted Conception

Marcia C. Inhorn, Rosario Ceballo and Robert Nachtigall

Introduction

For most Americans, the term 'infertility' conjures up images of wealthy, white, well-educated couples seeking high-tech, medical interventions. A skewed master narrative dominates perceptions of infertility in the US, such that views of infertility are deeply inflected by race and socio-economic filters. Medical and social science scholars have helped to maintain and perpetuate this dominant narrative by conducting research with the most readily accessible study populations: namely, white, economically privileged couples attending infertility treatment clinics (Abbey et al, 1991). This is true even among anthropologists who are concerned with issues of social and cultural diversity. Of the five book-length ethnographies devoted to infertility and assisted conception in the US, all have documented the struggles of white professional couples to make the 'elusive embryo' (Greil, 1991; Sandelowski, 1993; Becker, 1997, 2000; Thompson, 2005).

As a result, ethnic minorities' experiences of infertility and infertility treatment – including their attempts to access assisted reproductive technologies (ARTs) – are almost entirely missing from the social scientific literature. Seline Szkupinski Quiroga's (2007; n.d.) recent anthropological work is exceptional in this regard; she has documented the infertility and ART experiences of ethnically diverse couples in California, including the explicit fears of 'race mixing' that colour the practices of third-party gamete donation and surrogacy among physician providers.

Given the invisibility of ethnic minorities in the US infertility literature, we offer this chapter as a corrective. First, we attempt to theorize infertility as a potent form of 'health disparity' in the US: to wit, ethnic minorities suffer from higher rates of infertility, but receive fewer diagnostic and treatment resources. As such, minority infertility serves as a case *par excellence* of 'stratified reproduction', or the ways in which fertility is differently valued according to one's power and privilege within mainstream US society (Ginsburg and Rapp, 1995). We then proceed to the findings of our own studies of infertility among three marginalized ethnic minority populations – namely, African Americans (conducted by Ceballo), Latino/a Americans (conducted by Nachtigall and colleagues) and Arab Americans (conducted by Inhorn). As we will argue, these ethnic minority groups continue to be *marginalized* in mainstream US society. Their infertility and suffering remain *invisible*, including to social scientists. And the barriers they face in accessing appropriate infertility care index the very *unwantedness* of more 'black and brown babies' in a post-9/11 setting where xenophobia and anti-immigrant sentiment remain high, including in contemporary political discourse.

Infertility and health disparities

As is now recognized by the US National Institutes of Health (NIH), infertility prevalence rates are part and parcel of the overall picture of health disparities that continue to plague ethnic minority populations in the US.[1] In America, couples who are less well educated, lower income and from underprivileged minority groups are most likely to struggle with infertility (Aral and Cates, 1983; Scritchfield, 1995; Greil, 1997; Chandra and Stephen, 1998; Meyer, 1999; Molock, 1999; Jain and Hornstein, 2005). Why? Researchers have pointed to a variety of occupational hazards, environmental risks, and lifestyle factors, including smoking and obesity, that may predispose poor, ethnic minority popu-lations to infertility (Harris, 2006; Inhorn and Fakih, 2006). In addition, delays in treatment of reproductive tract infections – including those that are sexually transmitted and those that result from poor-quality medical care – may predis-pose minority women to reproductive problems that can result in infertility, including pelvic inflammatory disease and ectopic pregnancy (Molock, 1999).

Unfortunately, those most likely to be infertile in the US are also those least likely to seek highly specialized medical services. Among infertile women, acquiring medical services is positively associated with older age, a college education, a high income and having a Euro-American white racial background (Stephen and Chandra, 2000; Feinberg et al, 2007). Even in Massachusetts, a state with mandated comprehensive insurance coverage for infertility and assisted conception, couples who seek medical care for infertility tend to be highly educated, of upper socio-economic status, and from white, Euro-American backgrounds (Jain and Hornstein, 2005). For instance, among 561 women who attended an infertility clinic at Brigham & Women's Hospital in Boston, Massachusetts, more than 60 per cent of the patients had household incomes over $100,000, compared with only 18 per cent in the state's general population (Jain and Hornstein, 2005).

Economic barriers to access are a major part of this story of underutilization. More than 420 in vitro fertilization (IVF) clinics exist in the US alone, and 107,000 ART procedures give rise to 40,000 IVF babies annually (Centers for Disease Control and Prevention, 2004). However, in the US in 2002, the mean cost per IVF cycle was estimated at $9547 – in a country where the gross national income per capita in the same year was only $33,360 (Jain, 2006). Moreover, this does not account for the fact that there is wide variation in ART pricing *between* clinics, many of which charge well over $10,000, and as much as $20,000, per ART cycle in the US today. While such costs are increasingly being covered by the US insurance industry, thereby allowing some middle class and even working class couples to access ARTs (King and Meyer, 1997), many insurance plans provide only minimal coverage for IVF and related services. Although insurance plans may pay for the costs of blood work and ultrasounds, the expensive hormonal medications and the ART procedures themselves may not be covered under most health insurance policies. In short, because of the lack of universal insurance coverage in the US, ARTs remain a private, fee-for-service form of healthcare delivery in most American states, accessible largely to white, middle to upper class infertile couples.

Furthermore, as ARTs have come to dominate reproductive medicine and to dictate practice trends, attention to low-technology infertility treatment has all but disappeared in the US. As a result, low-income infertile women and men face increasing difficulty in obtaining even the most basic diagnostic services (Becker, 2000). Access to infertility treatment services, even 'low-tech' ones, may be limited or even non-existent (Zambrana, 1987; Rodin and Ickovic, 1990; Inhorn and Fakih, 2006). As noted by Nsiah-Jefferson and Hall (1989, p95),

> *Members of minority communities have an equal or even greater need for programs to treat infertility, but ... these needs have not been defined as a legitimate concern and ... treatments are generally not available to low-income women, who are disproportionately nonwhite. Going beyond this clear mismatch between the needs and services available, the ... issue for low-income women and women of color comes down to the social construction of infertility as a 'social problem.' Why have the infertility problems of minority communities been ignored? What are the implications for the daily life and social status of low-income women and women of color?*

These questions point to a potent ideological issue – namely, the existence in the US of what some have called 'stratified reproduction' (Ginsburg and Rapp, 1995) and others have called a 'eugenic logic of IVF' (Steinberg, 1997); to wit, ARTs are being used to enhance the fertility of married white elites, thereby producing 'white babies for married couples who are able to pay for them' (Nsiah-Jefferson and Hall, 1989). In the white majority view, infertility is seen as a 'non-issue' for low-income and minority couples, who are seen as being 'hyper-fertile' and undeserving of further children (Roberts, 1997; Ceballo, 1999; Inhorn and Fakih, 2006). Indeed, the fact that infertility treatment services are largely restricted to white elite couples in the US provides a salient example of stratified reproduction, or 'the arrangements by which some reproductive futures are valued while others are despised' (Ginsburg and Rapp, 1995, p3).

For African Americans in particular, negative stereotypes about black women's sexuality carry deeply entrenched historical roots (Roberts, 1997; Collins, 2000). Consequently, tenacious racial stereotypes play a significant role in many African-American women's experiences with infertility. Historically, African-American women were exploited economically for their labour as slaves, as well as sexually exploited and raped by white men (King, 1988). Collins (2000) posits that negative stereotypes about female slaves' hypersexuality and hyperfertility provided a justification for their sexual exploitation. African-American women today live with these lingering stereotypes in the form of negative public images that characterize their sexuality and reproductive abilities as flawed, irresponsible and dangerous. Social stereotypes of African-American women depict them as sexually promiscuous or aggressive; overly fecund and irresponsibly having too many babies; and neglectful, lazy mothers (Collins, 2000). Indeed, in *Killing the Black Body*, Dorothy Roberts (1997) contends that many public policies have deliberately punished African-American women precisely *for* having children.

It is not surprising, then, that public images of infertility hardly ever include African-American women, or other women of colour (e.g. Latinas, Native Americans, Arab-American Muslim women) (Quiroga, 2007). Their 'invisibility' in the world of American infertility – in infertility clinics, in infertility support groups, in media stories and in infertility scholarship – bespeaks a very potent form of stratified reproduction (Ginsburg and Rapp, 1995), indeed, outright racism in American society.

The consequences of this invisibility and marginalization may be devastating for minority infertile couples, especially those coming from subcultural groups where parenthood (and particularly motherhood) is glorified and children are highly valued, including for their tangible contributions to family life. On a social and cultural level, many minority groups in the US can be described as 'pronatalist', in that children and parenthood are seen as desirable social attributes. Couples may be expected to have children early within marriage, and may have larger numbers of children in the household and younger age structures than the US population at large (Inhorn and Fakih, 2006). Children may be perceived as a source of social status, power and immortality, particularly when family structures are patriarchal and patrilineal (Inhorn and Fakih, 2006). Thus infertility may precipitate a social crisis for both men and women in childless marriages, whose very social identities are determined by their ability to reproduce. Although women may carry the greatest social burden of infertility, in terms of blame for the reproductive failing, marital duress and social ostracism, men may also suffer over their own infertility and childlessness, particularly because infertility and impotency are popularly conflated, and paternity may be seen as bound to manhood among ethnic minority populations (Lloyd, 1996; Webb and Daniluk, 1999; Inhorn, 2002, 2003a, 2003b, 2004; Inhorn and Fakih, 2006). Among these subcultural groups, crises of infertility may be handled through religiously based coping. For example, many researchers have illuminated the central role played by the church and Christian religious participation in negotiating health problems in the African-American community (Chatters and

Taylor, 1989; Levin et al, 1995; Chatters, 2000; Mattis, 2002). Similarly, Islam provides solace for many Muslim couples facing the crisis of infertility (Inhorn 1994, 1996, 2003a; Inhorn and Fakih 2006).

Research settings, methods and study populations

Given this context, this chapter seeks to examine the experiences of infertility and assisted conception among three minority ethnic groups, living in two major urban centers in the US (metropolitan Detroit, Michigan, and the San Francisco Bay Area, California). In Michigan, African Americans were contacted primarily through the obstetrics and gynaecology department of a large university teaching hospital, while Arab Americans were contacted through a private, Arab-serving IVF satellite clinic in the ethnic 'enclave' community of Dearborn. In California, Latinas and Latinos were contacted while attending the infertility clinic of a large, university-affiliated county general hospital.

In each case, in-depth, semi-structured, and, in most cases, tape-recorded interviews were conducted with 50 African-American women, 88 Latina women and 33 Latino men, and 23 Arab-American women and 40 Arab-American men, for a total of 234 individuals (161 women, 73 men).[2] It is important to note that most of the respondents were married, although many of the interviews were conducted with individuals alone. The goals of the research were to explore the experiences of infertility, the impact on social and psychological well-being, the use of treatment services and the coping strategies used to face this life problem.

Most of the research subjects did not have higher degrees, and some had not completed high school, particularly those who had emigrated to the US from Latin America and the Middle East. Conditions of poverty were common throughout the study sample, reflecting national patterns. Even though a significant percentage of African Americans, Latinos/as, and Arab Americans have achieved middle class status or higher in the US, most members of these groups are lower income, with many families existing below the national poverty line (Wayne State University, 2004).

It is also important to note that in Michigan, African Americans and Arab Americans have been severely affected in recent years by changes in the urban industrial workforce, particularly the 'slow death' of the US automobile industry in cities like Detroit and Flint, Michigan. The closing of factories and the outsourcing of US manufacturing jobs to foreign countries has dealt a devastating blow to the Michigan economy, particularly for poor and minority urban families. In this study, many infertile couples were struggling to make ends meet, and, by 2007, some who had been gainfully employed in the auto industry had lost their jobs and were existing on unemployment compensation. Similarly, in California, Latino immigrants tend to occupy the lowest economic rungs of society, working primarily in low-paying service, domestic, construction and agricultural sectors. Furthermore, US immigration policies militate against full citizenship rights for many Latinos, who exist on the margins of American society as 'illegal aliens'. Among all of these minority populations, many families lack health insurance, with negative implications for health and well-being.

Indeed, economic impoverishment and accompanying low social class status are major problems for all three of these ethnic minority populations within the landscape of America. Poverty affects the ability of African Americans, Latinos/as and Arab Americans to seek higher education, improve their standard of living and access affordable healthcare, including for problems of infertility, as we shall see.

African Americans

Many obstacles hinder African-American couples from seeking medical care for infertility, including economic disadvantage, an absence of referrals, lack of knowledge about services and treatments, poor insurance coverage, a variety of cultural sanctions and a healthy suspicion of the medical establishment based on long histories of discriminatory treatment (Caesar and Williams, 2002). African-American women who do seek medical treatment tend to have experienced difficulty conceiving for longer periods of time compared to their white counterparts (Jain, 2006). Furthermore, infertility clinics are typically selective about whom they will treat, limiting services to married, heterosexual couples who can afford private, fee-for-service healthcare.

For many of the African-American women in this study, fertility was initially assumed and taken for granted; thus the presence of an infertility problem was typically experienced as a highly stressful, devastating life event. Most of the women experienced infertility as a trauma that tore at the very foundations of their sense of self and womanhood. Anxious to overcome the problem, 90 per cent of the African-American women in the study had discussed their infertility with a physician. However, numerous barriers prevented the majority from undergoing any form of assisted conception. These included the high costs of medical treatments, lack of knowledge about such treatments, lack of access to knowledgeable infertility specialists, and husbands' refusal to participate or to continue seeking medical interventions.

Financial barriers were particularly prominent for African-American couples in the study. One woman, who had been trying to become pregnant for three years, discussed the perceived financial burden of infertility treatment by saying:

> *White women are like – 'Look, me and my husband, we're going to get a thirty thousand (dollar) loan, and we'll get the IVF and we'll have kids.' But black people are like, 'Look, I can't afford that. I just barely have health insurance.'*

Another woman explained:

> *If they could look and tell me what was wrong, I didn't have any money for them to fix it. So it didn't matter.*

Some women in the study attempted to 'negotiate' with their healthcare providers in order to afford some form of treatment. Creative solutions included buying hormonal medications on a daily basis, or 'on instalment'. One woman, who was denied the possibility of purchasing 'by the vial', poignantly described the experience of holding a full box of expensive medication in her hands before having to return it. She compared the experience to:

> ... *giving up every chance I ever had of having a baby. I felt like I was giving up a baby, giving it back. And I was walking really slow. I was really depressed. Just walking down the hall and trying not to cry in front of anybody ... I gave it to them real slowly. I said, 'I'm sorry I can't afford it', you know, and I felt bad. And he [the doctor] felt bad.*

Many African-American women in the study struggled with their infertility in silence and isolation. While reproductive difficulties are often an extremely private matter for people of any race, the depth of the loneliness among infertile African-American women in this study had an important racial and cultural component. Black women's solitary experiences with infertility are in keeping with social stereotypes about African-American women's sexuality, as well as public images of the typical white infertile American couple. Such images are directly reinforced in some US infertility treatment settings. For example, in viewing the publicly posted photos of IVF babies at a hospital's infertility clinic, one African-American woman explained:

> *I saw all these pictures of these babies, but I didn't hardly see any black babies and I thought, 'These doctors are helping people have babies, and are they just like picking certain people to help?' And I also thought, 'These people don't want to see any black babies.'*

Moreover, some of the African-American women in this study had internalized the racial stereotypes they had faced as African-American women. A few women thought that they were the 'only one' – 'the only black woman walking the face of the earth who cannot have a baby' (Ceballo, 1999). One woman compared herself to other black women, saying,

> *For so many of my sisters, it's just not a problem getting pregnant, you know? The problem is trying not to get pregnant. And that was just so hard to deal with.*

Another African-American woman had come to the conclusion that infertility must be 'a white thing'.

As a result, a few of the African-American women who participated in this study had never before discussed their feelings and experiences with anyone outside of their close family members, choosing instead out of shame or embarrassment to struggle through the pain in isolation. One woman in the study agreed to be interviewed because she endured 11 years of infertility by herself. She declared:

> *I will share my story with you because there wasn't anybody for me to share [with] … There wasn't anybody black to share with me. There wasn't anybody that understood.*

Many of the women believed that no one else could understand their experiences, unless they, too, had been infertile. As one woman explained, 'I had a wall up to everyone else.' Not surprisingly, the pain of infertility strained and shattered relationships, sometimes with husbands, other family members and close friends.

In order to endure the pain and isolation of infertility, African-American women in the study tended to cope through religiosity and spirituality, which were often described as an indispensable part of their lives. Many of the women in this study found solace in viewing infertility as part of God's plan. For instance, one woman described how she:

> *… prayed a lot and thought that what we were doing was the right thing … that having a family is something we wanted and that it was smiled upon by God. I would always pray little prayers, like throughout the day, especially if I was feeling really discouraged. And that was helpful. It brought me a sense of peace.*

Yet religious faith was not a completely unambiguous presence in these women's lives. Some women expressed a more complex and ambivalent relationship with God and their church communities. For example, in response to her infertility, people often reminded one woman in the study that, 'God works in mysterious ways'. Her reaction to this proverb was that:

> *God works in mysterious ways, but I also know that if you don't have an egg, okay, well, I don't think I'm Mary. I don't think that it's going to happen. I am mad with God, and I think I'm being punished.*

Latinos/as

Whereas African-American women in the study belonged to a variety of Protestant Christian denominations, more than 85 per cent of the Latino/a women and men in the study were Catholic, and they subscribed to the belief that it is a duty to God to have children. Among Latinos, both women and men believed that a child was the basis of the marital relationship: a child was thought to create a bond between the couple and legitimize the relationship. Given the centrality of children in the couple relationship, almost everyone in the Latino study began trying to conceive as soon as they decided that they were in a relationship, often before marriage. Slightly less than half of those interviewed were living together but were not married; as they explained, they first wanted to test their fertility together.

When conception did not occur, infertility was a devastating experience for Latino couples, with far-reaching effects on women and men individually, as well as on the couple. Latino cultural assumptions about men's and women's roles underpinned the marital dynamics of couples unable to conceive. Childless marriages were considered a failure, and there was a widespread expectation that the relationship would end if no children were born. Even long-term relationships were threatened by infertility, as women and men became demoralized by repeated failures to conceive and increased fault-finding with each other. There was a high degree of marital conflict because of childlessness, often revolving around who was to blame for the infertility.

Because parenthood continues to be a strong cultural expectation for Latinos, and childless families are viewed as incomplete, women and men faced derision and public scrutiny. As a result, the gender identity of both women and men was undermined. The inability to become a mother was at odds with basic Latino philosophies about womanhood, such as *marianismo,* or the idea that a woman's self-esteem is manifested in her ability to be a generous mother and maintain strong traditions of family (Comas-Diaz, 1989). Latina girls continue to be socialized for marriage and childrearing to the exclusion of work-related or school-related roles (East, 1998). Therefore, unwanted childlessness leaves a void for most Latina women that nothing else can fill.

Recent work suggests that Latino men have much broader roles than stereotypes of machismo have portrayed (Gutmann, 1996; Torres et al, 2002). Yet the presence of infertility in the couple relationship was a considerable threat to men's masculinity, even if they did not themselves have an infertility factor. The stigma of male infertility was even greater than for female infertility. Men who did not have children were thought to have compromised sexuality, and men reported that they had been denigrated by others, through questions such as, 'What's wrong? Can't you do it?' Given the emphasis on virility in notions of machismo (Torres, 1998), and the confusion that men in the general US population often have between virility and potency (Nachtigall et al, 1992), the finding of a low sperm count was deemed a great threat to men's masculinity. Not surprisingly, women in the study reported that men were often unwilling to submit to a semen analysis. Infertility is thus a challenge to Latino male role expectations, and a source of culturally induced shame and anger, as has been found for men in the general infertility population (Nachtigall et al, 1992), as well as in other cultures (Inhorn, 2002, 2003a, 2003b, 2004).

Given this pressure to conceive, it is no surprise that both men and women vowed to persevere until they conceived a child. Yet, as with the infertile African Americans described above, low-income Latinos in the study faced a number of challenges that impeded their ability to receive medical treatment for infertility. Over half of this sample reported years of trying to conceive before seeking medical intervention. Reasons given for not seeking care sooner included: lack of economic resources; belated awareness of existing resources; the conviction that difficulty conceiving is a private matter; and the belief that conception may take time and perseverance will pay off.

Furthermore, a primary impediment to treatment seeking was communication. Language and cultural barriers resulted in patients having difficulty both in understanding diagnoses and treatments, and in communicating their questions, concerns and experiences to physicians. For example, Latina women in this study did not want to be called 'infertile', or to have their problem referred to as 'infertility'. They believed that the term meant that there was no hope to have a child, and that the label would jinx their efforts to conceive.

As described earlier, all of the Latinos/as in this study were interviewed at an infertility clinic in a large public teaching hospital. Unfortunately, an arcane and disorganized hospital bureaucracy resulted in patients having difficulty with appointment scheduling, follow-up visits and timed laboratory procedures. Furthermore, as with the African-American population, the low-income Latino men and women in this study had limited economic resources, and could not pursue medical treatment beyond the most basic level.

An interesting difference emerged between Latina women raised in the US and women who had emigrated to the US as adults. Women raised in the US considered healthcare to be a right, were more assertive in seeking infertility care, and had an overall plan that they intended to pursue in order to become pregnant. In contrast, immigrant women expressed more gratitude for the care they received, took less initiative in considering other biomedical possibilities, and most did not have a plan that went beyond what the clinic could offer them. When couples had exhausted the treatment available at this clinic, only a few had other economic resources to pursue additional treatment. For example, only one couple in the study had undergone IVF, in this case in Mexico. Only a few couples had any additional financial resources with which to pursue private medical treatment.

As a result of these many barriers to effective infertility care, women in this study tended to rely on traditional infertility treatments, sometimes simultaneously with biomedicine. Humoral medicine was the most popular option. Used in many parts of the world (e.g. Inhorn, 1994), humoral medicine relies on a system of oppositions, such that problems thought to be caused by 'cold' must be treated through 'hot' remedies. The primary infertility-related diagnosis Latina women received from traditional practitioners was 'cold womb'. The primary treatment for cold womb was massage, followed by herbal medicine, usually in the form of a hot tea, accompanied by staying warm, usually at home.

The majority of women in this study reported that they had been to a masseuse, or *sobadora*, at some point in their quest for conception, often while seeking biomedical care. Sobadoras are common throughout Mexico and Central America. Some women reported that their mothers took them for massage when they reached puberty, to ascertain if they were in good reproductive health. Such *sobadoras* also practise in Latino communities in the US; for example, several could be found near the infertility clinic site.

Latina women in this study generally attempted to find solutions to their infertility that were familiar and within their economic means. The inability of this population to afford assisted conception furthered feelings of desperation and efforts to try anything. Women in particular found it hard to imagine a life without children. They reported that they intended to continue trying to conceive until they reached menopause. Some women in the study who were in their 40s had been trying to conceive for more than 20 years. Only a few women had jobs that they found interesting, and no one thought a good job would be an acceptable alternative to having children. Every woman continued to hope for conception and reported feeling desperate at the idea of never conceiving. Women also questioned whether their relationships could succeed without a child.

Interestingly, Latino men were more willing than women to consider adoption. The majority of women felt that to try to adopt was to give up hope of having a biological child. They were also concerned that their extended families would not accept an adopted child. However, after years of no conception, some women in the study had considered adoption as they gave up hope of ever having a biological child. Unfortunately, most couples in the study could not afford to adopt, either in the US or abroad. They did not have the economic means, as adoption can cost even more than assisted conception. Adoption in countries of origin was not deemed possible without considerable funds to underwrite the expense.

Arab Americans

As in the Latino/a study, Arab Americans can be described as pronatalist, in that children and parenthood are seen as desirable, even necessary social attributes. Couples of Middle Eastern descent living in the US are expected to have children early within marriage, as reflected in the relatively high marriage and fertility rates in this population (Schopmeyer, 2000). For Arab Americans, children are a source of social status. For Arab-American men in particular, social power is achieved in patriarchal, patrilineal family structures through the birth of children, especially sons, who will perpetuate patrilineal structures into the future (Inhorn, 1996). Thus infertility may precipitate a social crisis for both men and women in childless marriages, whose very social identities are determined by their ability to reproduce.

As with African Americans and Latinos, Arab Americans may turn to their religion, usually Islam, to make sense of their suffering. The Islamic scriptures describe infertility as a God-given condition, thereby providing a satisfying religious reason for why some individuals are infertile (Inhorn, 1994, 2003). However, the Islamic scriptures also disallow alternative modes of family formation, including legal adoption (Serour, 1996; Bargach, 2002). Assisted conception with third parties (i.e. gamete donation and surrogacy) is also disallowed within the dominant Sunni branch of Islam (Serour, 1996; Inhorn, 2003a, b). Thus, unlike other infertile couples in the US, who may resort to donor insemination and adoption to overcome their childlessness, infertile Arab-American Muslim couples generally have no other way of becoming parents except through medical treatment, which is encouraged in Islam as a religious obligation (Kulwicki, 1996).

As with African-American and Latino couples, Arab Americans may have great difficulty accessing affordable infertility care. Most of the Arab-American couples in the study were either economic immigrants from Yemen, who had come to Michigan to work in the auto industry, or political refugees from Lebanon and Iraq, who had experienced war and persecution in their home countries. Few of these individuals were fluent in English, and some were illiterate in both English and Arabic. In virtually all cases, Arab Americans in the study had sought out diagnostic and treatment services from a fellow Arab-American (Muslim) physician, which whom they could feel cultural and linguistic rapport as fellow Arabs and fellow Muslims. Nonetheless, most of the men and women in the study lacked Western understandings of reproductive physiology, which represented major barriers to negotiating infertility care. Furthermore, many Arab-American women (and their husbands) were uncomfortable receiving gynaecological care from a male physician, because of cultural notions of modesty and shame. Thus, the clinic made extra attempts to employ female support staff.

As with the Latina study population, poverty was the major deterrent to receiving effective infertility care among the Arab Americans in the study. The majority of women did not work, partly because of cultural expectations, but also because of lack of English and driving skills. Most of the men in the study were working in low-wage, blue-collar or service sector occupations, mainly as gas station attendants, dishwashers and busboys in Middle Eastern restaurants, truck drivers, construction workers, auto mechanics or store clerks. Thus salaries were generally low in this study population, with many couples 'eking out' subsistence lives.

Furthermore, most Arab-American couples did not have private health insurance that would cover the costs of infertility diagnosis and treatment (in a state that does not mandate insurance coverage for infertility services). Most couples did not own credit cards. As a result, virtually all of their financial transactions in the infertility clinic setting were handled in cash, which was exchanged over the counter at the end of clinic visits.

Relatively few couples in the study were able to pay for infertility treatments, especially ARTs. Indeed, for many, the economic barriers to seeking ARTs were insurmountable, even with discounts that were sometimes offered out of sympathy. As a result, only seven intracytoplasmic sperm injection (ICSI) cycles had been completed among the 50 couples in the Arab-American study, and only two cycles had resulted in living offspring. The vast majority of couples reported no spontaneous pregnancies and no initiation of ART cycles, despite more than five years of marriage in most cases.

Most couples in the study were deeply demoralized, because conceiving a biological child was unlikely without some sort of financial miracle. As one man explained his situation:

> *Money is the problem. If anybody who is infertile can afford to do it [ARTs], he would do it. But even if we need [ARTs], we cannot afford to do it right now. The doctor gave us a discount and said it will cost only $7000. But I don't have even $100. What can we do?*

Such economic barriers were true even among those who had received advanced educations in their home countries and were able to speak English fluently. Education and English fluency were no guarantee of good jobs and financial security, particularly for those who had come to the US as political refugees.

In short, Arab-American couples in this study uniformly desired children, were concerned about the future of their marriages in an ethnic enclave community where marital fertility was expected and scrutinized, and were willing and encouraged by their religion to undertake biomedical treatments in order to solve their infertility problems. Yet, as Muslims, none of them could accept donor technologies as an option, and very few would contemplate adoption, which is explicitly forbidden in the Islamic scriptures. Severe economic constraints impinged upon their abilities to seek even low-tech forms of infertility care, a problem that they routinely lamented in interviews. Indeed, most couples in the study cited the high cost of treatment in the US as their main barrier to care, and many of them had at least contemplated returning to the Middle East, where the cost of a single ART cycle is generally less than $3000 (or, in some countries, may be subsidized by the state). However, for most, even the cost of a return trip to the Middle East was prohibitively expensive, and many couples refused to return home (generally to Iraq or Lebanon) out of fear for their lives.

Arab-American men and women often remarked that they could never afford to undertake assisted conception, either in the US or abroad, without borrowing large sums of money from friends, family or a bank. The few couples who had undertaken ARTs had usually received such a loan in order to subsidize a single cycle. When the cycle failed to lead to conception, they described their shock and demoralization. For example, one young Iraqi couple, who had saved up enough money to try one cycle of IVF, were faced with the loss of the husband's auto shop job at the same time that the wife experienced a serious IVF complication (i.e. ovarian hyperstimulation syndrome). She was hospitalized for a week without health insurance. At the time of the interview, the couple was praying that a local Islamic charity would cover the $30,000 hospital bill. The wife explained:

> *This is the worst time ever in both of our lives. I never got that much sick before. This is the most toughest time I've ever had.*

In general, Arab Americans in this study described their lives as 'hard' and 'stressful', given the traumatic conditions that had led them to flee their home countries and their inability to return safely home because of ongoing political violence. Once in America, they had also faced problems of economic hardship, exclusion and discrimination. Most of the Arab-American couples in the study were 'unassimilated' into US society at large, were unable to speak English fluently and were unable to mingle freely outside the protective enclave of their ethnic community.

It is important to note that a long history of racial discrimination, negative stereotyping and hate crimes against Arab Americans have been documented by researchers (Suleiman, 1999; Leonard, 2003; McCloud, 2003). However, the events of 11 September 2001 have clearly reversed the generally assimilationist efforts of Arab Americans to 'blend' into white US society as an 'invisible' (and racially unmarked) ethnic minority population (Naber, 2000). Today, 'Arabs' are vilified by many white Americans, who regard Arab men as particularly dangerous, untrustworthy, inherently violent and religiously fanatical. These caricatures of

Arab-American men also include images of male hypersexuality and hyperfertility; to wit, Arab-American men and Muslim men in general are seen as polygamous fathers of children from multiple wives, harking back to Western Orientalist fantasies of the harem (Said, 1978). If Arab-American men are portrayed as hyperfertile polygamists in the Western popular imagination, then the very possibility that they might suffer from real infertility problems within stable, loving, monogamous marital unions may lead to the convenient denial of their legitimate infertility problems. Indeed, in the post-9/11 landscape, research on Arab-American health has been sidetracked, resulting in 'a critical need for Arab Americans and the research community to take up the challenge of actively developing and funding research, education, and intervention programs' (Hassoun, 1999, p174).

Conclusion

As shown in this comparison of infertility among African Americans, Latinos/as and Arab Americans in the US, all three ethnic minority populations experience the effects of stratified reproduction, including poverty, lack of access to afford- able, high quality reproductive healthcare, and marginalization within US society as a whole. Sadly, all of these populations share in common their poor health status *and* the combination of fear and prejudice displayed by many white Americans. All of these populations face significant reproductive disruptions, but are despised as reproducers in a racist/classist/xenophobic society. Only with further studies of infertility among ethnic minority populations in the US can we begin to combat these forms of stratified reproduction and to shed light on the very humanity and dignity with which members of these oppressed minority groups strive to overcome the many barriers that face them – including those that prevent the infertile from becoming loving parents.

Notes

1   A scientific workshop, entitled 'Health Disparities in Infertility', was held on 10–11 March 2005, at the National Institutes of Health in Bethesda. Several of the papers from the workshop were subsequently published in *Fertility and Sterility*, vol 85, no 4, April 2006.
2   In the Arab-American study, most participants were uncomfortable being tape-recorded. Thus, handwritten notes were taken and transcribed following interviews.

References

Abbey, A., Andrews, F. M. and Halman, L. J. (1991) 'Gender's role in responses to infer-tility', *Psychology of Women Quarterly*, vol 15, pp295–316

Aral, S. O. and Cates, W. (1983) 'The increasing concern with infertility: Why now?', *JAMA: Journal of the American Medical Association*, vol 250, no 17, pp2327–2331

Bargach, J. (2002) *Orphans of Islam: Family, Abandonment, and Secret Adoption in Morocco*, Rowman Littlefield, Oxford

Becker, G. (1997) *Healing the Infertile Family: Strengthening Your Relationship in the Search for Parenthood*, University of California Press, Berkeley, CA

Becker, G. (2000) *The Elusive Embryo: How Women and Men Approach New Reproductive Technologies*, University of California Press, Berkeley, CA

Caesar, L. G. and Williams, D. R. (2002) 'Socioculture and the delivery of health care: Who gets what and why', *The ASHA Leader*, vol 7, pp6–8

Ceballo, R. (1999) '"The only black woman walking the face of the earth who cannot have a baby": Two women's stories', in M. Romero and A. J. Stewart (eds) *Women's Untold Stories: Breaking Silence, Talking Back, Voicing Complexity*, Routledge, New York

Centers for Disease Control and Prevention (2004) *Assisted Reproductive Technology Success Rates: National Summary and Fertility Clinic Reports, 2001*, Centers for Disease Control and Prevention, Atlanta, GA

Chandra, A. and Stephen, E. H. (1998) 'Impaired fecundity in the United States: 1982–1995', *Family Planning Perspectives*, vol 30, pp34–42

Chatters, L. M. (2000) 'Religion and health: Public health research and practice', *Annual Review of Public Health*, vol 21, pp335–367

Chatters, L. M. and Taylor, R. J. (1989) 'Age differences in religious participation among black adults', *Journal of Gerontology*, vol 44, no 5, ppS183–S189

Collins, P. H. (2000) *Black Feminist Thought: Knowledge, Consciousness, and the Politics of Empowerment*, 2nd edition, Routledge, New York

Comas-Diaz, L. (1989) 'Culturally relevant issues and treatment implications for Hispanics', in *Crossing Cultures in Mental Health*, Society for International Education Training and Research, Washington, DC

East, P. L. (1998) 'Racial and ethnic differences in girls' sexual, marital, and birth expectations', *Journal of Marriage and the Family*, vol 60, pp150–162

Feinberg, E. C., Larsen, F. W., Wah, R. M., Alvero, R. J. and Armstrong, A. Y. (2007) 'Economics may not explain Hispanic underutilization of assisted reproductive technology services', *Fertility and Sterility*, vol 88, pp1439–1440

Ginsburg, F. D. and Rapp, R. (1995) *Conceiving the New World Order: The Global Politics of Reproduction*, University of California Press, Berkeley, CA

Greil, A. L. (1991) *Not Yet Pregnant: Infertile Couples in Contemporary America*, Rutgers University Press, New Brunswick, NJ

Greil, A. L. (1997) 'Infertility and psychological distress: A critical review of the literature', *Social Science & Medicine*, vol 45, no 11, pp1679–1704

Gutmann, M. C. (1996) *The Meanings of Macho: Being a Man in Mexico City*, University of California Press, Berkeley, CA

Harris, L. (2006) 'Challenging conception: A clinical and cultural history of in vitro fertilization in the United States', PhD thesis, University of Michigan, Ann Arbor, MI

Hassoun, R. (1999) 'Arab-American health and the process of coming to America: Lessons from the metropolitan Detroit area', in M. Suleiman (ed) *Arabs in America: Building a New Future*, Temple University Press, Philadelphia, PA

Inhorn, M. C. (1994) *Quest for Conception: Gender, Infertility, and Egyptian Medical Traditions*, University of Pennsylvania Press, Philadelphia, PA

Inhorn, M. C. (1996) *Infertility and Patriarchy: The Cultural Politics of Gender and Family Life in Egypt*, University of Pennsylvania Press, Philadelphia, PA

Inhorn, M. C. (2002) 'Sexuality, masculinity, and infertility in Egypt: Potent troubles in the marital and medical encounters', *Journal of Men's Studies*, vol 10, pp343–359

Inhorn, M. C. (2003a) *Local Babies, Global Science: Gender, Religion, and In Vitro Fertilization in Egypt*, Routledge, New York

Inhorn, M. C. (2003b) '"The worms are weak": Male infertility and patriarchal paradoxes in Egypt', *Men and Masculinities*, vol 5, pp238–258

Inhorn, M. C. (2004) 'Middle Eastern masculinities in the age of new reproductive technologies: Male infertility and stigma in Egypt and Lebanon', *Medical Anthropology
Quarterly*, vol 18, pp34–54

Inhorn, M. C. and Fakih, M. H. (2006) 'Arab Americans, African Americans, and infertility: Barriers to reproduction and medical care', *Fertility and Sterility*, vol 85, no 4,
pp844–852

Jain, T. (2006) 'Socioeconomic and racial disparities among infertility patients seeking
care', *Fertility and Sterility*, vol 85, no 4, pp876–881

Jain, T. and Hornstein, M. D. (2005) 'Disparities in access to infertility services in a state
with mandated insurance coverage', *Fertility and Sterility*, vol 84, no 1, pp221–223

King, D. K. (1988) 'Multiple jeopardy, multiple consciousness: The context of a Black
feminist ideology', *Signs*, vol 14, no 1, pp42–72

King, L. and Meyer, M. H. (1997) 'The politics of reproductive benefits: US insurance
coverage of contraceptive and infertility treatments', *Gender & Society*, vol 11, pp8–30

Kulwicki, A. (1996) 'Health issues among Arab Muslim families', in B. Aswad and B.
Bilge (eds) *Family and Gender among American Muslims: Issues Facing Middle Eastern
Immigrants and their Descendants*, Temple University Press, Philadelphia, PA

Leonard, K. I. (2003) *Muslims in the United States: The State of Research*, Russell Sage
Foundation, New York

Levin, J. S., Taylor, R. J. and Chatters, L. M. (1995) 'A multidimensional measure of religious involvement for African Americans', *The Sociological Quarterly*, vol 36, no 1,
pp157–173

Lloyd, M. (1996) 'Condemned to be meaningful: Non-response in studies of men and
infertility', *Sociology of Health and Illness*, vol 18, pp433–454

Mattis, J. S. (2002) 'Religion and spirituality in the meaning-making and coping experiences of African American women: A qualitative analysis', *Psychology of Women
Quarterly*, vol 26, pp308–320

McCloud, B. (2003) 'Islam in America: The mosaic', in Y. Y. Haddad, J. I. Smith and J.
L. Esposito (eds), *Religion and Immigration: Christian, Jewish, and Muslim Experiences
in the United States*, AltaMira, Walnut Creek, CA

Meyer, C. S. (1999) 'Family focus or career focus: Controlling for infertility', *Social
Science & Medicine*, vol 49, pp1615–1622

Molock, S. D. (1999) 'Racial, cultural, and religious issues in infertility counseling', in L.
H. Burns and S. N. Covington (eds) *Infertility Counseling: A Comprehensive Handbook
for Clinicians*, Parthenon Publishing, New York

Naber, N. (2000) 'Ambiguous insiders: An investigation of Arab American invisibility',
Ethnic and Racial Studies, vol 23, pp37–61

Nachtigall, R. D., Becker, G. and Wozny, M. (1992) 'The effects of gender-specific diagnosis on men's and women's response to infertility', *Fertility and Sterility*, vol 57,
pp113–121

Nsiah-Jefferson, L. and Hall, E. J. (1989) 'Reproductive technology: Perspectives and
implications for low-income women and women of color', in K. S. Ratcliff (ed)
Healing Technology: Feminist Perspectives, University of Michigan Press, Ann Arbor, MI

Quiroga, S. S. (2007) 'Blood is thicker than water: Policing donor insemination and the
reproduction of whiteness', *Hypatia*, vol 22, pp143–161

Quiroga, S. S. (n.d.) 'Disclosure, social support, and identity work among infertile
couples', unpublished manuscript

Roberts, D. (1997) *Killing the Black Body: Race, Reproduction, and the Meaning of Liberty*,
Vintage, New York

Rodin, J. and Ickovic, J. R. (1990) 'Women's health: Review and research agenda as we
approach the 21st century', *American Psychology*, vol 45, pp1018–1034

Said, E. (1978) *Orientalism*, Vintage, New York

Sandelowski, M. (1993) *With Child in Mind: Studies of the Personal Encounter with Infertility*, University of Pennsylvania Press, Philadelphia, PA

Schopmeyer, K. (2000) 'A demographic portrait of Arab Detroit', in N. Abraham and A. Shryock (eds) *Arab Detroit: From Margin to Mainstream*, Wayne State University Press, Detroit, MI

Scritchfield, S. A. (1995) 'The social construction of infertility: From private matter to social concern', in J. Best (ed) *Images of Issues: Typifying Contemporary Social Problems*, 2nd edition, Aldine de Gruyter, Hawthorne, NY

Serour, G. I. (1996) 'Bioethics in reproductive health: A Muslim's perspective', *Middle East Fertility Society Journal*, vol 1, pp30–35

Steinberg, D. (1997) 'A most selective practice: The eugenic logics of IVF', *Women's Studies International Forum*, vol 20, pp33–48

Stephen, E. H. and Chandra, A. (2000) 'Use of infertility services in the United States: 1995', *Family Planning Perspectives*, vol 32, pp132–137

Suleiman, M. W. (ed) (1999) *Arabs in America: Building a New Future*, Temple University Press, Philadelphia, PA

Thompson, C. (2005) *Making Parents: The Ontological Choreography of Reproductive Technologies*, MIT Press, Cambridge, MA

Torres, J. B. (1998) 'Masculinity and gender roles among Puerto Rican men: A dilemma for Puerto Rican men's personal identity', *American Journal of Orthopsychiatry*, vol 68, pp16–26

Torres, J. B., Solberg, V. S. H. and Carlstrom, A. H. (2002) 'The myth of sameness among Latino men and their machismo', *American Journal of Orthopsychiatry*, vol 72, pp163–181

Wayne State University College of Urban, Labor and Metropolitan Affairs (2004) 'Arab, Chaldean, and Middle Eastern children and families in the tri-county area', *Detroit Metropolitan Census 2000*, Fact Sheets Series 4, February, pp1–30

Webb, R. E. and Daniluk, J. C. (1999) 'The end of the line: Infertile men's experiences of being unable to produce a child', *Men and Masculinities*, vol 2, pp6–25

Zambrana, R. E. (1987) 'A research agenda on issues affecting poor and minority women: A model for understanding their health needs', *Women & Health*, Winter, pp137–160

Glossary

- **ART**
  Assisted reproductive technology(ies). A collective term for specialist infertility treatments in which gametes are manipulated outside of the body. The aim of these treatments is the improved chance of pregnancy. Examples of ART include IVF and ICSI.
- **Cycle**
  One complete attempt with a treatment; e.g. a cycle of IVF.
- **Donor**
  Person providing their gametes (eggs or sperm) for use by another person or couple.
- **Donor insemination**
  Placing sperm from a donor (i.e. not the woman's partner) into the reproductive tract of a woman to produce a pregnancy.
- **Embryo**
  The term for a fertilized egg at up to eight weeks' development.
- **Embryo transfer (ET)**
  The transfer of one or more embryos to the uterus after in vitro fertilization.
- **Endometriosis**
  A condition in which endometrial tissue (that which usually lines the womb) is found outside of the womb.
- **Gamete**
  The female egg or male sperm.
- **Hysterosalpingography**
  An x-ray of the uterus and fallopian tubes that involves the injection of dye through the cervix, used to diagnose tubal blockage.
- **Infertility**
  A term most frequently used to describe the inability to achieve a pregnancy after 12 months of unprotected sexual intercourse (WHO). Infertility can be attributed to the following diagnostic categories: male factor infertility (including problems of sperm quantity, quality or motility); and female factor infertility (including ovulatory problems, tubal blockage, **endometriosis** and problems with the uterus). In a proportion of cases, there is a combination of these factors inhibiting pregnancy. A couple can also be diagnosed as having **unexplained infertility**, in which no discernable cause can be found. Prevalence rates for each of these categories of infertility vary considerably by country.

- **In vitro fertilization (IVF)**
  A technique in which fertilization takes place outside of the body in a petri dish. IVF can be carried out using the eggs from the female partner, which are then fertilized using her partner's sperm, or by using **donor** sperm. **Donor** eggs can also be used, fertilized with either the male partner's sperm or donated sperm, meaning that a resulting child can be genetically related to both partners, just the female, just the male, or neither partner.
- **Intracytoplasmic sperm injection (ICSI)**
  A technique in which an individual sperm is injected directly into a single egg in an attempt to achieve fertilization. The procedure for assisted conception is then very similar to that followed in conventional **IVF**. ICSI has significantly improved the 'treatment' of male infertility and is especially useful where donor sperm is unacceptable.
- **Intrauterine insemination (IUI) or artificial insemination (AI)**
  A technique in which a sample of prepared sperm is inserted into the cervix or uterus of a woman. The procedure is often accompanied by ovarian stimulation in order to increase the chances of pregnancy.
  IUI or AI with **donor** sperm (DI) is often used when there are problems with the male partner's sperm or if there is a risk that he may pass on an inherited disease.
- **Laparoscopy**
  Investigation of the abdomen using key hole surgery techniques.
- **Polycystic ovarian syndrome**
  A hormonal problem in which multiple cysts form on the ovaries. Symptoms include infertility, absent or infrequent periods, acne, obesity and hirsutism.
- **Pre-Implantation Genetic Diagnosis (PGD)**
  A medical procedure in which embryos created through IVF are screened in order to select out those with a particular disease or condition. PGD is considered to be a controversial technology since it allows the possibility of screening for 'social' attributes, such as gender and conditions not deemed to be medically related or life threatening.
- **Premature menopause / premature ovarian failure**
  Menopause starting at a much earlier age than is considered usual.
- **Surrogacy**
  A method of reproduction whereby a woman agrees to become pregnant and deliver a child for a contracted party. In traditional or conventional surrogacy the woman who carries the pregnancy is the child's genetic mother, conceiving through insemination with the infertile woman's partner's sperm or donor sperm. In gestational surrogacy, the surrogate conceives using an embryo from the infertile woman and her partner (or from an embryo donor) and carries the baby on behalf of the infertile couple.
- **Unexplained infertility**
  A term for infertility where there is no obvious medical cause.

Index

.